Time, Temporality and Global Politics

EDITED BY

ANDREW HOM, CHRISTOPHER MCINTOSH,
ALASDAIR MCKAY & LIAM STOCKDALE

E-INTERNATIONAL RELATIONS PUBLISHING

E-International Relations
www.E-IR.info
Bristol, England
2016

ISBN 978-1-910814-15-4 (Paperback)
ISBN 978-1-910814-16-1 (e-book)

Production: Michael Tang
Cover Image: milosluz

A catalogue record for this book is available from the British Library.

E-IR Edited Collections

Series Editors: Stephen McGlinchey, Marianna Karakoulaki and Agnieszka Pikulicka-Wilczewska

Editorial assistance: Paul Blamire, Matthew Koo and Jan Tattenberg

E-IR's Edited Collections are open access scholarly books presented in a format that preferences brevity and accessibility while retaining academic conventions. Each book is available in print and e-book, and is published under a Creative Commons CC BY-NC 4.0 license. As E-International Relations is committed to open access in the fullest sense, free electronic versions of all of our books, including this one, are available on the E-International Relations website.

Find out more at: http://www.e-ir.info/publications

Recent titles

Environment, Climate Change and International Relations

Ukraine and Russia: People, Politics, Propaganda and Perspectives

System, Society and the World: Exploring the English School of International Relations (Second Edition)

Restoring Indigenous Self-Determination

Nations under God: The Geopolitics of Faith in the Twenty-first Century

Popular Culture and World Politics: Theories, Methods, Pedagogies

Caliphates and Islamic Global Politics

About the E-International Relations website

E-International Relations (www.E-IR.info) is the world's leading open access website for students and scholars of international politics, reaching over 250,000 unique visitors each month. E-IR's daily publications feature expert articles, blogs, reviews and interviews – as well as student learning resources. The website is run by a registered non-profit organisation based in Bristol, England and staffed with an all-volunteer team of students and scholars.

Abstract

International Relations scholars have traditionally expressed little direct interest in addressing time and temporality. Yet, assumptions about temporality are at the core of many theories of world politics and time is a crucial component of the human condition and our social reality. Today, a small but emerging strand of literature has emerged to meet questions concerning time and temporality and its relationship to International Relations head on. This volume provides a platform to continue this work.

The chapters in this book address subjects such as identity, terrorism, war, gender relations, global ethics and governance in order to demonstrate how focusing on the temporal aspects of such phenomena can enhance our understanding of the world.

About the Editors

Andrew Hom is a postdoctoral fellow in Politics at the University of Glasgow's School of Social and Political Sciences, researching the temporality of victory in war. His research interests include timing and time, security, international theory, the vocation of IR, and philosophies of social science. Previously, he taught at the University of St Andrews and Vanderbilt University, after earning a PhD from Aberystwyth University and sundry degrees from the University of Kansas. Examples of his work can be found in *Security Dialogue*, *Review of International Studies*, *International Studies Review* and *Military Review*.

Alasdair McKay is an editor-at-large at E-International Relations. He holds undergraduate and postgraduate degrees in politics from the universities of Manchester and Aberystwyth. He has worked for an African human rights NGO and in the parliamentary office of an MP.

Christopher McIntosh is a currently a Visiting Assistant Professor of Political Studies at Bard College. He has broad research interests in international relations theory and security studies, including: time/temporality in IR theory and practice; the relationship between the concepts of war and sovereignty and the implications this has for the contemporary practice of political violence; U.S. strategy in the war on terror; and the likelihood of nuclear terrorism. His work has been published in *International Theory, Studies in Conflict and Terrorism* and *Orbis*. He is currently working on a book project on how the concept of time informs IR scholarship.

Liam Stockdale is a postdoctoral fellow at the Institute on Globalisation and the Human Condition and the Institute for Innovation and Excellence in Teaching and Learning at McMaster University, Hamilton, Canada. His research interests sit at the intersection of international relations theory, critical security studies and contemporary social and political thought – with a particular emphasis on the temporalities of societal governance. He is the author of the book *Taming an Uncertain Future: Temporality, Sovereignty, and the Politics of Anticipatory Governance* (Rowman & Littlefield, 2016), and has previously published on the relationship between time and globalisation, and the global politics of sport.

Contents

Introduction

ALASDAIR MCKAY

E-INTERNATIONAL RELATIONS, UK

Adolf Hitler once said that 'time in this war, as in all historical process, is not a factor valuable in itself but must be weighed up' (quoted in Maiolo 2013: 229). He further opined that time 'will work against us if we do not use it properly' (quoted in Gellately 2007: 375). As early as 1937, Hitler believed that success hinged upon acting at the right time (see Faber 2009). For Hitler, quick decisive victories were often seen as preferable to long campaigns because he understood that the longer the war went on the harder it would be for the Germans to maintain military advantages (Fischer 2011: 144-45). Blitzkrieg (lightning war) was deployed to ensure the speedy victories he desired. The defeat of France was made possible by Germany's quick intensity of warfare and the failure of the French military to understand 'the quickened rhythm of the times' that was behind German military strategic thinking during that campaign (Bloch 1968: 45). America's increasing support for the UK meant that if Hitler was to realise his plan of creating an Eastern Lebensraum, he needed to do it quickly (Fritz 2011: 39). By 1945, when defeat was looming, Hitler lamented that 'the tragedy for us Germans is that we never have enough time' (quoted in Fischer 2011: 145).

This example drawn from the Second World War gives a flavour of how time is relevant to major international events. Yet, out of all the books and journals that have been published under the rubric of International Relations (IR) very few have paused to ponder the role of time or temporality. Time has often occupied the role of an entity that sits in the background whilst events transpire in the foreground and it essentially 'functions as context or as analytical indicator, but not as a distinct component deserving judicious investigation' (Hom 2008: 2). As Kimberly Hutchings (2008: 11) observes: the study of world politics has traditionally been 'overtly pre-occupied with spatial rather than temporal relations'. The pre-occupation with space has meant that there has been 'little space for time' (Hom 2010: 1) in analyses of world affairs from IR scholars.

In recent years some IR scholars have turned their attention to time and the temporal dimension of international politics. Writers have shown how the adoption of what could be described as a 'temporal lens' (see Hom 2008; Stockdale 2013) can greater enhance our understanding of various human phenomena. This book seeks to serve as a vessel for this blossoming literature.

Time, Temporality and the Social Sciences

For humans, time has an ambiguous and perhaps paradoxical quality to it. In some ways, it is something that we seem to push to the back of our thoughts in the same way a timepiece sits unthreateningly on the walls; it is 'simply what the mechanical clock and Gregorian calendar display, a neutral and enumerated dimension in which life unfolds' (Hom 2013: 2). Yet, it is also a mysterious concept that has always slipped into the human mind's ideas about change, impermanence, and mortality.

Humans are, so far as we know, the only animal with a capacity to reflect on past events and envision the future. We can, unlike other birds and beasts, project ourselves in time (see Routledge and Arndt 2005). The passing of time is accompanied by the human observation of change; this can be seen by noticing the falling of leaves in autumn, our own bodily changes, or the growth and development of others. Whilst time can be affiliated with creation, birth, and change for the better, it can also bring destabilising change, decay and death. 'Time' is very much, as Schwartz wrote, 'the school in which we learn' and 'the fire in which we burn' (Schwartz 1968).

Everything we do is embedded in time and we are in some way fundamentally aware of time, yet there is something about time that makes it beyond the human mind's capacity to fully grasp. The mathematician Alfred Whitehead (1919: 73) once stated 'it is impossible to meditate on time and the mystery of the creative passage of nature without an overwhelming emotion at the limitations of human intelligence'. Centuries earlier, St. Augustine (2008: 217) famously made a similar point with his musings on the question 'what is time?' – 'If no one asks me, I know; if I wish to explain it to one that asketh, I know not'. We seem to think we know what time is, yet it seems difficult to precisely express what we mean when we speak of it. Despite time being one of the most commonly used words in speech – often we describe how time 'flies', 'slowed down', is 'running out', was 'wasted' or 'killed' – there is not a universally agreed upon understanding of time. Indeed, throughout history, time has been understood and explained in multiple ways. This includes moving in a linear direction; cyclical repetitive change; as foundational, functional, social and artistic (McCullough 1991:1); as a measure such as seconds, minutes and hours; or as lived experience or a form of social regulation, such as 'clock time' or 'Western standard time' (see Gell 1992, 1998; Adam 1994, 1995, 2002, 2006; Hughes and Trautmann 1995; Aveni 2000; Urry 2000, Hom 2010). It is a subject that has served as a muse for poets, novelists, musicians, artists and filmmakers. There has also been considerable engagement with time and related concepts in philosophy, mathematics, the human sciences, the physical sciences and the social sciences.

Social scientists have not shied away from confronting these issues. As Helga Nowotny (2015:4) observes, 'libraries are full of detailed and also of general investigations on social science and time'. Major thinkers of their times such as Emile Durkheim, George Herbert Mead, Marcel Mauss, Henri Hubert, and Lewis Mumford all theorised on time from various angles. In more recent times, this tradition has been continued by the likes of Eviatar Zerubavel, Georges Gurvitch, Karl Mannheim, Julius A. Roth, Alfred Schutz, Ptirim Sorokin & Robert K. Merton, Norbert Elias, Niklas Luhmann, Michel Foucault, Anthony Giddens, Barbara Adam, Helga Nowotny, Elizabeth Grosz, Bob Jessop and Patrick Baert (see Ryan 2004 for an introductory overview of time and social theory). Recent work on time and related issues from scholars has led some to identify a 'temporal turn' in the social sciences (Hassan 2010). This renewed interest in time can be seen as a reaction to the 'spatial turn', that emerged in the late-1970s where globalisation was examined predominantly through reference to changes in the spatial contours of existence. Undoubtedly, the spatial aspect of globalisation cannot be ignored, but time, it is argued, is equally important in this process because changes in the temporality of human activity inevitably generate altered experiences of space or territory.

With reference to the temporal aspect of globalisation, some have focused their work on the fragmentation of 'linear, irreversible, measurable, predictable time' and the emergence of a detemporalised life in 'timeless time' (Castells 2000: 429; see also 1998); the increased internalisation of time in 'the brains and bodies of the citizens' (Hardt and Negri 2000: 23) or the multitemporal multiscalar nature of globalisation (Sassen 1992; 1994; 1999; 2000). But, perhaps some of the most influential and consistent theorising on globalisation has identified the process as embodying a marked shift in the perceived acceleration of social life. Marxist geographer David Harvey, who interestingly enough criticised other scholars for adopting a dialectical approach that privileges time and ignores space (1996: 4), described this phenomenon as the 'time space compression' (Harvey 1989: 240). For Harvey, contemporary developments in capitalism, information, and communications technologies have led to the speeding up of the circulation of capital and a perceived speeding up of social life in general which is simultaneously reducing the significance of space.

The perception that life is speeding up has been a subject of much discussion for social scientists in the past two decades. The crux of this line of thinking is that the rhythms of life and the social and cultural change, aided by the invention and spread of technologies—such as mobile phones, personal computers, and the Internet—have radically shortened spatial and temporal distances. Scholars have used different terms to describe this perceived social acceleration and disagree about the exact sources of these recent shifts in the spatial and temporal contours of human life. Nevertheless, there is some consensus that alterations in humanity's experiences of space and time are working to erode the importance of local and even national boundaries in many arenas of human civilisation. In the spirit of those who had warned about what this development might entail for humanity, such as Virilio (1977) and McLuhan (1964), a significant section of literature has examined the political, cultural and ecological consequences of this development (Scheuerman 2004; Hassan and Purser 2007; Hope 2011; Bastian 2012; Hassan 2012; Rosa 2013; Sharma 2014; Wajcman 2015).

Whether or not one accepts that a temporal turn in the social sciences has occurred, IR has, until recently, not treated time with anything like the same level of interest as other schools of the social sciences or academic disciplines in general. IR as a whole exhibits something of a 'temporal blindness' (Stockdale 2013, 5) compared to other disciplines.

IR Scholars and Time

IR scholars are relatively new to theorising on time and placing it at the foreground of analysis. As this book will show, scholars are now writing explicitly about time and its relationship to world politics. Still, there are earlier works to consider. James Der Derian's writings on diplomacy (1992) and war (2001), that owed a considerable intellectual debt to Virilio's theorisation of speed, endeavoured to focus on the temporal aspect of politics. Much of this work focuses on the collapse of the importance of space created by the previously mentioned perceived increase in the pace of life, and the political ramifications of this development. Within this perspective, the increase in the pace of the modern world—witnessed in all sorts of areas such as in transport, weapons, media—has resulted in a collapse of distance. This influences relations between states because what counts more and more in their strategic relations is the speed of travel, of weapons, of information and so on. Consequently, the political control and management of time is becoming more important than the control and distribution of territory, meaning that 'international relations is shifting from a realm defined by sovereign places, impermeable borders and rigid geopolitics, to a site of accelerating flows, contested borders and fluid chronopolitics' (Der Derian

1992: 129-130). This is what Der Derian (1990) refers to as the '(s)pace problem in international relations' – 'pace' becomes more important than 'space.'

The influence of Der Derian's writing can be seen in Michael J. Shapiro's work on temporality, the state and citizenship. Understanding citizenship as not simply a spatial but also a temporal phenomenon, Shapiro (2000) examined the consequences of the perceived quickened pace of life for the concept of citizenship and its relationship to the nation-state. Similar themes can be seen in the work of R.B.J. Walker (1991, 1993), who was for a while one of the few IR theorists to look at time in relation to state sovereignty. Essentially much of Walker's work is designed to 'draw attention to the contradictory, antagonistic, aporetic or radically undecidable character of the most consequential principles enabling modern political life, especially in relation to prevailing accounts of state sovereignty and its limits' (Walker 2010, 16). To Walker, the temporal aspect plays a pivotal role in driving this contradictory and antagonistic character. His work *Inside/Outside* (1993), whilst not devoted entirely to time, presented an analysis of problems over the concept of state sovereignty caused by the increasing significance of 'the experience of temporality, of speed, velocity and acceleration' (1993: 5) in modern life. For Walker, adequately understanding this development requires an examination of the role of not only space, but also time. Indeed, 'conceptions of space and time' he writes 'cannot be treated as some uniform background noise, as abstract ontological conditions to be acknowledged and then ignored' (Walker 1993, 130-131). In Walker's argument, the principle of state sovereignty is first of all a spatial resolution of the relationship between universality and particularity (Walker 1993: 11, 78, 177), but, importantly, it also serves as a temporal resolution. This is because inside the state there is a story of time as linear progress, which makes it possible for universalist aspirations such as perpetual peace and prosperity to come true. Outside the state, time is seen as cyclical repetition of conflicts and wars.

State sovereignty attempts to reconcile time and space, but with 'the startling velocity of contemporary accelerations' (Walker 1993: 178) comes the reality that temporality can no longer be contained within spatial boundaries. Indeed, 'the hope that temporality may be tamed within the territorial spaces of sovereign states alone is visibly evaporating' (Walker 1993: 155). In Walker's words, the acceleration of modern life has caused the distinction between time inside the state and outside the state to collapse. Today's world is a time of speed and acceleration, in which space is compressed and borders are becoming less significant, and so too is state sovereignty. For Walker, the brightest future for global politics lies in reframing it in a new light that avoids this paradoxical conflict between the national and the international, which would involve redefining the concepts that we use to explore the nature of political relations.

Elsewhere in the literature, *Time and Revolution* by Stephen Hanson (1997) argued that the history of Marxism and Leninism reveals an unsuccessful revolutionary effort to reorder the human relationship with time and that this failure had a direct impact on the design of the political, socioeconomic, and cultural institutions of the Soviet Union. *Trauma and the Memory of Politics* by Jenny Edkins (2003) explored the remembrance of traumatic events such as war, genocide and terrorist attacks, and how these seek to serve as methods to reinforce feelings of nationhood – and yet also challenge it. Through several case studies, employing various scholarly approaches, she illustrated that some forms of trauma remembrance focus on the horror of the traumatic event and harness its memory to promote change and to challenge the political systems that initially spawned the violence of wars and genocides. William Callahan (2006) examined time and its role in national identity formation in international society through his work on China's 'National Humiliation Day'—a special holiday declared by the head of state during wartime which is celebrated in local churches throughout the country. In *Times of Terror* (2008), Lee Jarvis noted that narratives, such as claims to temporal discontinuity, linearity and timelessness, were all central to the narration of the George W. Bush administration's War on Terror.

Kimberly Hutchings' *Time and World Politics* (2008) was the first work in IR dedicated entirely to time and its relationship with global politics. She investigated the thoughts about time which inhabit Western political philosophies, illuminating the significance of the ancient Greek concepts Chronos and Kairos in intellectual thought and political life. Chronos is understood as a medium through which we can measure lifespans, periods and empires which ensures the certainty of death and decay. It comprehends time as predictable, inevitable and in a sense 'normal'. Kairos (meaning the right or opportune moment), is the 'transformational time of action, in which the certainty of death and decay is challenged' (Hutchings 2008: 5). It is 'exceptional time', affording a sense of human agency. Hutchings argues chronotic time makes it simple to view life, and politics, as ticking along to the rhythms of a linear, unimpeded process of history. But, granting a role for Kairos in politics makes the flow of Chronos uncertain and problematic because through its ascription of human agency, it is essentially struggling with Chronos' seemingly inevitable course and steering it in a different direction. She later shows that this struggle between Chronos and Kairos applies to both international theorists and political actors, illustrating them to be tormented by the idea that politics is a project of controlling time as Chronos and creating a different kind of time through Kairos. 'World political time', to Hutchings (2008: 21), represents an effort to balance these two concepts into a singular understanding of time.

The Time behind IR

Despite the neglect of time in the discipline, IR theories are heavily anchored by temporal assumptions. Generally, there are two main templates for understanding temporality that underpin the discipline – the cyclical and the linear. For those unfamiliar with such terms, it is worth briefly discussing them. Cyclical theories of time are influenced largely by classical cosmology, where, drawing upon observations of nature, all aspects of the world are temporally organised in a rotational pattern of the seasons, tides, menstrual cycles, birth and death, rise and fall, growth and decay, and structured in relation to the rotation of the planets (Adam 2004: 18). It has been noted that in many ancient societies historical time was construed as 'cyclical', where time was generally represented as a wheel, whose revolutions symbolised the circularity of history playing out in an endlessly recurrent pattern (see Rosen 2014).

In IR, realists generally posit a cyclical view of history. They conceive international relations as a 'realm of recurrence and repetition' (Wight 1966: 26), where self-interested states are 'doomed to repeat the behaviour appropriate to rational actors with differing capabilities in an anarchic context' (Hutchings 2008: 13). Neorealists do not explicitly assume that human nature is evil, but they do posit that the only rational behaviour for states in an anarchical international system is to assume that all other states are potentially aggressive and must be defended against, a reality that they see continually repeating itself (Kaufman 2014: 2). In sum, realism implies a static temporality because 'the substance of nations' relentless struggle for power— their ascent, then decline, then eclipse in international politics—remains essentially the same' (Kaufman 1996: 349).

Linear time tells a different story to cyclical time as it sees time flowing in a particular direction and following a teleological trend. This more modern account of temporality has two main historical intellectual influences. The first is the rise of Christian historiography which, drawing on the teleological representations of temporality in the Judeo-Christian tradition, construed history as a sequence of events unfolding in a linear fashion. Events in biblical scripture clearly indicate a unidirectional notion of time moving from Creation to Apocalypse. The consequence of this understanding was that episodes in life were understood as having a beginning, middle and end and this ordering of experience had 'the potential to redeem human existence from the futility of nature's endless repetition' (Schaap 2005: 83).

The second intellectual resource is the historical progressivism of the European Enlightenment, which relies on an inference of the future from the known present or past which gave rise to a sequential understanding of historical development. No longer perceived as reliant on divine will, human destiny now lay in the hands of humans themselves—who could guide it towards a victory for reason and rationality. Modernity was and still is, 'revolutionary time', because it is 'the time in which progress through human intervention is possible, if not inevitable' (Hutchings 2008: 112). Within this type of thinking, then, there is also a belief in a moral direction to human history (Russell 1990:143–45; see also Hom and Steele 2010). Rousseau, Hegel and Marx expressed this idea in different ways but nonetheless argued for a dialectical progression towards the universal, though they diverged on what form the universal would take (Hom and Steele 2010, 276). Nevertheless, they all believed in a universalising aspect of time's arrow.

Linear time displaced cyclical time in societies and is essentially the dominant method for how we understand and measure time today. Hom (2010) has chronicled the rise of this conception of time to the status of the 'hegemonic metronome'. This regime of time's ascension went hand in hand with the rise of territorial sovereignty. Together these two drove to standardise a precise and coordinated measurement of time which 'buttressed the edifice of political modernity' (Ibid: 1156) as it spread globally. Western standard time, once established, was then exported to other civilisations. Hom (2010: 1168) concludes that Western standard time constitutes 'modernity's most global hegemon' having achieved an almost unquestioned position as the dominant method of time measurement cross-culturally (see also Attrill 2013: 6-35).

The concept of linear time flows through much writing from the traditions of liberalism and neoliberalism in IR. Liberals tend to see the historical development of man's power and freedom as inevitable, irreversible, and achievable. Whilst liberals do not write explicitly about the perfectibility of the human form, they insist that the international system holds widespread opportunities for international cooperation (Kaufman 2014: 2). As Hutchings (2008: 13) observes, liberal thought exhibits a temporality of limited teleology where a more peaceful world, characterised by global democratisation, can be arrived at through international cooperation.

Whilst this line of thought is most common amongst contemporary liberals, they are not the sole heirs to this linear-progressive understanding of time. Some forms of constructivism lean towards embracing a progressive linear temporality. Wendt's (1992) belief in the inevitability of global government is an example of this as it assumes that history is essentially moving towards a world state (see also Hom 2008: 35). Some Critical Theories draw heavily on classical Marxism's views of history as a linear image of progress through stages of economic development from feudalism to capitalism, and then after revolution, on to communism. But other Critical Theories adopt a more pessimistic view of linear time that could be characterised as linear-regressive, seeing history as heading towards decline, because no feasible political alternative to liberal capitalism can be found. In particular, the works of Virilio and Agamben have been accused of leaving 'very little room for worldly hope' in their understandings of temporality (Hutchings 2008: 131).

Following the traditions of social theorists, some IR theorists have critiqued these dominant models of temporality and proposed alternative approaches to understanding time. For Hutchings, a key problem with our dominant understanding of time is that it is so committed to forging a singular understanding of time that it excludes any alternative accounts of temporality, a practice which has significant political consequences. Drawing upon postcolonial and feminist thinking, Hutchings argues that we should entertain the possibility of multiple, co-existing, and diverse visions of temporality, described as 'heterotemporality' (Hutchings 2008: 4) which would 'undermine the idea that we can theorise world-political time in homogeneous or unified terms' (Ibid: 155). Hom and Steele (2010) in their critiques of conceptions of time in IR have discussed an 'open time' that moves beyond the constraints of popular paradigms and 'encourages creativity while cautioning against a simulated, idealised self, thus assisting in the classically realist self-limitation of political action that preserves both present ethical concerns and the possibility of progress in the near and distant future.'

Accounting for the Neglect

The reasons for IR's neglect of time are difficult to identify. One could argue that the elusive, stealthy dynamic of time makes it hard to analyse in the same way as space. After all, we can observe the effects of time in the world, but there is a seemingly invisible quality to it and it is taken for granted as being represented by clock time. As Adam (1992: 175) observed of time as an analytical category for social scientists, it is 'deeply taken for granted in our daily lives and our social theories, it is not easily accessible to conscious reflection and social science analysis. This means that time needs to be made visible before its pervasive role in modernity can be appreciated'. This may provide a partial explanation but not a reliable excuse, because, after all, 'many IR scholars spend much of their working day analysing similarly invisible and imperceptible concepts, such as social structures, discourses, and identities' (Morini, 2012). Other reasons may be found in the history of IR as a discipline. Hutchings (2008: 11) writes that for much of the Cold War 'the space of international politics was thought of as frozen in time', which became reflected in academe. Assuming that inter-state relations would always consist of the balance of power where states would pursue self-interest in a timeless historical vacuum, IR theorists devoted many hours, days, and years developing models to support this, with neorealist texts leading the way. Theorists simply compared political units and tried to explain and predict political outcomes. This period has been described as the 'behaviouralist revolution' in which 'space gradually became privileged over time and context in analyses of world politics' (Vaughn-Williams 2005: 115). As an object of scholarly interest, time only became a real consideration for IR scholars when the Cold War ended, because, in the words of Walker (1993: 3), 'the demolition of the Berlin Wall may have signalled an opening across territorial space, but it equally signalled an awareness of temporal velocities and incongruities.'

According to Hutchings (2008:14) 'the themes of temporality and history have come centre stage in debates about world politics in International Relations since the end of the Cold War' (Hutchings 2008: 14). Perhaps the two largest issues were whether we had come to 'the end point of mankind's ideological evolution and the universalization of Western liberal democracy as the final form of human government' (Fukuyama 1989) or a world in which mankind was doomed to engage in an endless clash of civilisations (Huntington 1993, 1996). Yet, such writing has largely relied on assumptions about time and temporality without any real critical reflection on such concepts.

The so called 'historical turn' in IR, merits brief discussion here. This research enterprise in IR arguably stemmed from 'an emerging consensus that history is taken far more seriously within the discipline' (Vaughan-Williams 2005: 117). Over the past two or three decades, many scholars have attempted to historicise the concepts which underpin international relations and inject temporality into the study of historical processes. Whilst there has been important work conducted in this 'turn'—much writing in historical sociology has done a good job of critiquing the ahistorical assumptions underpinning realism and neorealism (Lundborg 2016: 18)—there has been little willingness from this intellectual turn to confront temporality or time directly.

As Stevens (2016: 37) observes, 'important though the renewed emphasis on history is in IR, other aspects of time and temporality should also be of interest to students of the international, as they are elsewhere in the social sciences and humanities'. Not only do the chapters in this book address many other aspects of time and temporality in relation to global politics, they also provide a more critical approach to how we understand, and use, such concepts.

The Book

Andrew Hom and Ty Solomon's opening chapter addresses a key concept in IR theory: the construction of identity. The authors suggest that the key to unlocking identity's temporal qualities can be found by unpacking the emotional elements of the human activity of timing.

Kimberly Hutchings' chapter explores the relationship between assumptions about time and temporality and global ethics. Hutchings shows how thought experiments tend to reflect assumptions that are specific to the time and place of the theorist.

Tim Luecke introduces a generational framework as a device to understand the relationship between temporality and international relations. He shows how a generational analysis can be applied and illustrates its analytical value through a discussion of his research into changes in Germany's culture and foreign policy from World War II to the present.

Tim Stevens' chapter draws attention to the nature of chronometric administration as a form of global governance. The chapter explores how global time is governed through an assemblage of institutions, norms, standards, multilateral agreements and technologies that produce the sociotechnical *chronos*, the dynamic and negotiated 'time of the world'.

The chapter from Robert Hassan explores how in the network society, politics takes place upon a radically different communicative basis but that we have so far failed to appreciate the actual extent of the transformation. The chapter discusses that in its communicative basis, digital networks, unlike analogue ones, have no equivalence in nature, and serve only to alienate people from democracy and political power.

Kevin K. Birth's chapter discusses the tensions generated by the global distribution of Western timekeeping and shows the relationship between the values celebrated in calendars and the rhetorical strategies used by nations in two very different international debates: climate change and the leap second.

Valerie Bryson's chapter focuses on gender relations, labour and time. Focusing on the access to leisure or discretionary time between genders, the economic and cultural assumptions that underpin the global market economy, and the devaluing of non-market activities, Bryson argues that such developments are having a damaging effect in many societies and argues that it is in the interests of us all to rethink our human relationship with time.

In his chapter, Christopher McIntosh seeks to illustrate some of the ways in which attention to the temporal dimension in IR can better inform our understanding of war. This chapter uses a temporal frame to sketch out some of the important aspects of the concept of war as it is produced and reproduced in IR scholarship and practice.

Shahzad Bashir discusses temporality through reference to the contemporary politics of Islam. To Bashir, temporality is a crucial issue for the study of Islam because the (erroneous) sense of homogeneity attributed to Islam stems from the attribution of a single history to Islam. The chapter demonstrates how ISIS has leveraged twenty-first century technology to create a shallow temporality through its vision of the Islamic past, which serves as a weapon against its detractors.

Kathryn Fisher's chapter explores how temporal assumptions position the 'homegrown' terrorist threat as a 'conditional self': an actor from within and without. To Fisher, signposts, such as 'domestic', 'international', and 'homegrown', generalise actors along sweeping categorical assumptions which enable an ongoing exclusion from full belonging and an increase in insecurity for many with no relation to terroristic violence.

Liam Stockdale's chapter explores how the post-9/11 profusion of pre-emptive rationalities has transformed the politics of (in)security into a politics of time, and how this has significant implications for the way political power is organised and exercised in the global security context. Stockdale fleshes out these concepts through an examination of the Obama administration's targeting of American citizens such as Anwar al-Awlaki in its drone warfare programme.

The concluding chapter from Lundborg and Holmqvist raises four topics of contemporary world politics to illustrate the constitutive impact of time and temporality on political discourse: the politics embedded in the United Nations system, the climate crisis, global information flows, and practices of war and 'counterterrorism'. By tracing the competing temporalities at work in these examples, the chapter provides an overview of the types of inquiry that a focus on time can open up for future research.

References

Adam, B. (1990) *Time and Social Theory*. First edition. Philadelphia: Temple University Press.

Adam, B. (1992) 'Modern Times: The Technology Connection and its Implications for Social Theory'. *Time and Society* 1: 175-192.

Adam, B. (1994) *Time and Social Theory*. Second edition. Cambridge: Polity Press.

Adam, B. (1995) *Timewatch: The Social Analysis of Time*. Cambridge: Polity Press.

Adam, B. (2002) 'The Gendered Time Politics of Globalisation: Of Shadowlands and Elusive Justice'. *Feminist Review* 70: 3–29.

Adam, B. (2004) *Time*. Cambridge: Polity.

Adam, B. (2006) 'Time'. *Theory, Culture and Society* 23(2–3): 119–126.

Augustine (2008) *The Confessions of St. Augustine*, Modern English Version, Revell

Attrill, N. (2013) *Temporal Sovereignty in Modern International Politics: The Contemporaneous Rise of Western Standard Time with Territorial Sovereignty, and the Significance of this Relationship to Sovereignty in Modern International Relations*. Master's Thesis, the Victoria University of Wellington.

Aveni, A. F. (2000) *Empires of Time: Calendars, Clocks and Cultures*. London: Tauris Park Paperbacks.

Bastian, M. (2012) 'Fatally Confused: Telling the Time in the Midst of Ecological Crises.' *Environmental Philosophy* 9(1): 23-48.

BBC Archive Written Document (1939) The Transcript of Neville Chamberlain's Declaration of War. Available online at: http://www.bbc.co.uk/archive/ww2outbreak/7957.shtml?page=txt

Bloch, M. (1968) [1946] *Strange Defeat: A Statement of Evidence Written in 1940*. New York, NY: W. W. Norton & Company.

Callahan, W. A. (2006) 'War, Shame, and Time: Pastoral Governance and National Identity in England and America.' *International Studies Quarterly* 50: 395-419.

Castells, M. (1998) *The Information Age: Economy, Society and Culture: The Rise of the Network Society*. Malden, MA: Blackwell Publishers.

Castells, M. (2000) *The Rise of the Network Society*. Oxford, UK: Blackwell Publishers; [1st edition 1996]

Edkins, J. (2003) *Trauma and the Memory of Politics*. Cambridge: Cambridge University Press.

Der Derian, J. (1990) 'The (S)pace of International Relations: Simulation, Surveillance, and Speed.'. *International Studies Quarterly* 34: 295-310.

Der Derian, J. (1992) *Antidiplomacy: Spies, Terror, Speed, and War*. London: Basil Blackwell.

Der Derian, J. (2001) *Virtuous War: Mapping the Military-Industrial-Media-Entertainment Network Boulder*. CO, Oxford: Westview Press.

Faber, D. (2009) *Munich, 1938: Appeasement and World War II*. Simon & Schuste.

Fischer, K. P. (2011) *Hitler and America*, University of Pennsylvania Press.

Fukuyama, F. (1989) 'The End of History?.' *The National Interest.* Summer.

Fukuyama, F. (1992) *The End of History and the Last Man*. Free Press.

Fritz, S.G. (2011*) Ostkrieg: Hitler's War of Extermination in the East*. Lexington, KY: The University Press of Kentucky.

Gell, A. (1992) *The Anthropology of Time: Cultural Constructions of Temporal Maps and Images*. Oxford, UK; Providence, RI: Berg.

Gell, A. (1998) 'Time and Social Anthropology'. *Senri Ethnol Stud.* 43: 9–24.

Gellately, R. J. (2007) *Lenin, Stalin and Hitler: The Age of Social Catastrophe.* New York: Alfred A. Knopf; London, Jonathan Cape.

Hanson, Sn E. (1997) *Time and Revolution: Marxism and the Design of Soviet Institutions.* Chapel Hill, NC: University of North Carolina Press.

Hardt, M. and Negri, A. (2001) *Empire.* Cambridge MA and London: Harvard University Press.

Harvey, D. (1989) *The Condition of Postmodernity.* Oxford: Blackwell.

Harvey, D. (1996) *Justice, Nature, & the Geography of Difference.* Oxford: Blackwell.

Hassan, R. and Purser, R. (2007) (eds.) *24/7: Time and temporality in the Network Society.* Stanford, Calif.: Stanford University Press.

Hassan, R. (2010) 'Globalisation and the "Temporal Turn"': Recent Trends and Issues in Time Studies.' *The Korean Journal of Policy Studies.* 25(2): 83–102.

Hassan, R. (2012) 'Time, Neoloberal Power and the Advent of Marx's 'Common Ruin' Thesis.' *Alternatives: Global, Local, Political* 37(4)

Hom, A.R. (2008) *Time and International Relations Theory.* PhD thesis, University of Kansas.

Hom, A. R. (2010) 'Hegemonic Metronome: the Ascendancy of Western Standard time.' *Review of International Studies*, 36, 1146.

Hom, A. R. and Brent J. Steele (2010) Open Horizons: The Temporal Visions of Reflexive Realism. *International Studies Review* 12: 271–300.

Hom, A. R. (2013) Reckoning Ruin: International Relations Theorising and the Problem of Time. PhD Thesis, Aberystwyth University.

Hope, W. (2011) 'Crisis of Temporalities: Global Capitalism After the 2007-08 Financial Collapse.' *Time and Society* 20(1).

Hughes, D., and Trautmann,T (eds.) (1995) *Time: Histories and Ethnologies*. Ann Arbor, MI: The University of Michigan Press.

Huntington, S. (1993), 'The Clash of Civilizations?' *Foreign Affairs*, vol. 72(3): 22–49.

Huntington, S. (1996) *The Clash of Civilizations and the Remaking of World Order*. New York: Simon & Schuster.

Hutchings, K. (2004) 'Global Civil Society: Thinking Politics and Progress'. In: Baker, G. and Chandler, D. (eds) *Global Civil Society: Contested Futures*. Routledge Advances in International Relations and Politics. London: Routledge. 130-148.

Hutchings, K. (2008) *Time and World Politics: Thinking the Present*. Manchester: Manchester University Press.

Kaufman, R.G. (1996) 'E. H. Carr, Winston Churchill, Reinhold Niebuhr, and Us: The Case for Principled Democratic Realism'. In: Frankel, B. *Roots of Realism; Realism: Restatements and Renewal.* Routledge. 314-345.

Kaufman, S.J. (2014) 'A Symbolic Politics Theory of War'. University of Delaware Paper Prepared for Presentation at ISAC-ISSS Conference Austin, Texas. November 14-16.

Le Goff, J. (1992 [1986]) *History and Memory*. Translated by Rendall, S. and Claman, E. New York

Lundborg, T. (2016) 'The Limits of Historical Sociology: Temporal Borders and the Reproduction of the "Modern" Political Present.' *European Journal of International Relations*. 22(1)

Maiolo, J. (2013) 'To Gamble All on a Single Throw': Neville Chamberlain and the Strategy of the Phoney War.' In: Baxter, C, Dockrill, M. and Hamilton, K. (eds) *Britain in Global Politics*. Basingstoke: Palgrave Macmillan. 220–41

McCullough, E. J. (1991) 'Introduction.' In: McCullough, E.J. and Calder, R.L. (eds) *Time as a Human Resource*. Calgary: University of Calgary Press.

McLuhan, M. (1964) *Understanding Media: The Extensions of Man*, New York: McGraw Hill.

Morini, D. (2012) 'International Relations and Time'. *E-International Relations*. Available online at: http://www.e-ir.info/2012/05/05/international-relations-and-time/

Nowotny, H. (2015) *Time: The Modern and Postmodern Experience*. Polity

Oxford English Corpus, 'The OEC: Facts about the language'. Available online at: http://www.oxforddictionaries.com/words/the-oec-facts-about-the-language

Rosa, H. (2013) *Social Acceleration: A New Theory of Modernity*. New York: Columbia University Press.

Rosen R. M. (2004) (ed.) *Time and Temporality in the Ancient World.* Philadelphia: University of Pennsylvania Museum of Archaeology of Anthropology.

Routledge, C. and Arndt, J. (2005) 'Time and Terror: Managing Temporal Consciousness and the Awareness of Mortality'. In: Strathman, A. and Joireman, J. (eds) *Understanding Behavior in the Context of Time: Theory, Research, and Application.* Mahwah, N.J: Lawrence Erlbaum. 59-84.

Russell, G. (1990) *Hans J. Morgenthau and the Ethics of American Statecraft.* Baton Rouge, LA: Louisiana State University Press.

Ryan, D. (2004). 'Time'. In: Ritzer, G. (ed.) *Encyclopedia of Social Theory.* SAGE.

Sassen, S. (1992) *The Global City: New York, London, Tokyo, New York.* Oxford: Oxford University Press.

Sassen, S. (1994) *Cities in a World Economy.* PineForge: Thousand Oaks.

Sassen, S. (1996) *Losing control? Sovereignty in an Age of Globalisation.* New York: Columbia University Press.

Sassen, S. (1999) 'The Spatiality and Temporality of Globalisation'. Annual Lecture of Global and World Cities Study Group, Loughborough University. (Edited from transcript and posted 30th November 2000).

Sassen, S. (2000) 'The Global City: The De-nationalizing of Time and Space'.

Paper presented at Conference on 'El Territori en Las Societat de les Xarxes. Dinámiques Territorialsi Organitzaciò Territorial'. Barcelona, 2-3 October.

Schaap, A. (2007) 'The Time of Reconciliation and the Space of Politics'. In: Veitch, S. (ed.) *Law and the Politics of Reconciliation*. Aldershot: Ashgate. 9-31.

Shaap, A. (2005) *Political Reconciliation*, London: Routledge.

Sharma, S. (2014) *In the Meantime: Temporality and Cultural Politics*. Durham, NC and London: Duke University Press.

Shapiro, M.J. (2000) 'National Times and Other Times: Re-Thinking Citizenship.' *Cultural Studies*. 14(1).

Scheuerman, W. (2004) *Liberal Democracy and the Social Acceleration of Time*. Baltimore: Johns Hopkins Press.

Schwartz, D. (1968) 'Calmly We Walk Through This April's Day'. *Selected Poems* (1938-1958). Summer Knowledge.

Stevens, T. (2016) *Cyber Security and the Politics of Time*. Cambridge: Cambridge University Press.

Stockdale, L.P.D. (2013) Governing the Future, Mastering Time: Temporality, Sovereignty, and the Pre-emptive Politics of (In)security. PhD Thesis, McMaster University.

Urry, J. (2000) *Sociology Beyond Societies: Mobilities for the Twenty First Century.* London, UK: Routledge.

Virilio, P. (1977) *Speed and Politics*. New York: Semiotext[e].

Vaughan-Williams, N. (2005) 'International Relations and the 'Problem of History''. Millennium: *Journal of International Studies* 34(1): 115-36.

Wajcman, J. (2014) *Pressed for Time: The Acceleration of Life in Digital Capitalism*. Chicago: The University of Chicago Press.

Walker, R.B.J. (1991) 'State Sovereignty and the Articulation of Political

Space/Time'. *Millennium* 20(3): 445-61.

Walker, R.B.J. (1993) *Inside/Outside: International Relations as Political Theory*. Cambridge: Cambridge University Press.

Walker, R.B.J. (2010) *After the Globe, Before the World.* London: Routledge.

Wendt, A. (1992) 'Anarchy is What States Make of It: The Social Construction of Power Politics'. *International Organization* 46(2): 391-425.

Whitehead, A. (1919) *The Concept of Nature*. Cambridge: Cambridge University Press; reissued Mineola, NY: Dover Publications, Inc. 2004.

Wight, M. (1960) 'Why is There No International Theory?' *International Relations* 2(1): 35-48.

1

Timing, Identity, and Emotion in International Relations

ANDREW R. HOM & TY SOLOMON
UNIVERSITY OF GLASGOW, UK

While time and temporality have been mostly neglected in social identity theory, the idea of timing is wholly absent in any substantive sense. It receives no mention in seminal critical works on identity in IR (e.g. Lapid and Kratochwil 1996) and even efforts specifically attuned to the interplay of identity, narrative, and temporality (Ringmar 1996; Hall 1999; Callahan 2006; Steele 2008),[1] do not engage with timing as a theoretical category or a core empirical phenomenon. This silence about timing echoes in broader International Relations (IR) scholarship otherwise concerned with time (e.g. Walker 1993; Hutchings 2008; Hom and Steele 2010; Berenskoetter 2011; Lundborg 2011; Stockdale 2013; McIntosh 2015) as well as in temporally sensitive identity scholarship outside IR (Campbell, O'Rourke, and Silverstein 2010).

To be clear, by timing, we mean much more than the intuitive or idiomatic sense in which one exclaims 'nice timing!' after an unexpected or intentional concurrence of two events that help things work out in a desired way. Referring to such occurrences as colloquial matters of timing implies that they require little further explication and, more importantly, that they hold little potential for advancing knowledge about social phenomena. Rather, we understand timing as a fundamental human activity, a basic means of synthesising pertinent changes for practical and political effect. In the course of this chapter, we argue that combining social identity theory with this sense

---

[1]   Ringmar (1996:125), Hall (1999:126), Callahan (2006:404), and Steele (2008:77–83) do mention timing but none engage it as a theoretical resource or conceptual category in its own right. In later work, Steele (2010 chp. 3) focuses on timing as a key component of *parrhesia*, speaking truth to power.

of timing holds significant benefits for IR: in general, a better understanding of time and its relationship to various political processes; in particular, a clearer formulation of the laborious and fraught work by which social selves come into being 'over time'.

Using Timing to Unlock Time and Identity

Although the absence of 'timing' in social identity research is understandable, it is also regrettable because a fuller and more explicit treatment of timing holds the potential to highlight important and interesting aspects of identity in international politics. In particular, it helps explain the desirous dimensions of identity, which in turn explain the central emotional element of timing as a basic human activity. By desire, we refer to the affective dimensions of subjectivity that stem from the subject's fundamental lack of foundations, which engender desire for a stable sense of self. This desire often constitutes the emotional 'grip' (Glynos 1999) factor that helps to account for subjects' identifications with narratives that offer a seemingly stable vision of 'fullness.' Taken together, emotion and timing help clarify the political implications of any temporal, constructed identity. Before discussing these issues, however, we need to explicate just what we mean by 'timing'.

Timing

Following Hom's (2013, 2016) recent work on time and IR theory and drawing on the social theory of Norbert Elias (2007), we deploy 'timing' not merely as an intuitive matter of when something happens but as a substantive theoretical concept in its own right that is distinct from and indeed analytically and practically prior to 'time' and 'temporality'. Elias (2007:38–39) locates 'time' and 'temporality' as linguistic artefacts that emerge to symbolise a dynamic and practical timing activity, which is a process of creative synthesis that proceeds by reference to a timing standard. For him, 'time' refers not to an existential or metaphysical thing, object, or dimension, but to a socially established relationship between two or more changing phenomena, one of which has been used as the frame of reference for organising the other(s). More specifically, timing proceeds by integrating and coordinating pertinent changes according to the chosen timing standard.[2]

---

[2]   Timing's analytical priority to 'time/temporality' complements Elias' historical account of the rise of social time consciousness. Elias' argument is that timing first emerged as social groups developed agricultural practices that benefited from coordination with seasonal variations. Agricultural civilisations subsequently attached religious significance to both recurrent and unique changes in the natural environment due to their impact on timing, which concomitantly became the concern of priesthoods (Elias 2007:42–44). As timing became necessary to societal survival and development, 'time'

Although Elias (2007:43–44) focuses primarily on repeatable and natural changes as the standards for successful human timing, any change continuum can provide a standard so long as social agents can use it to successfully synthesise the 'when-aspects' of various change continua they care about.[3] The timing standard does so by establishing priorities (which changes matter, and which of these matter more/most?), indicating how integration and coordination should proceed (what should happen and when?), and suggesting what compromises and sacrifices will be necessary. When successful, the timing standard organises changes in such a way that we are better able to orient ourselves in the (social) world, direct our individual and collective actions, and exercise self- and social control. When we have timed them successfully, we understand the relations between and importance of various changes and can decide how and when to act upon them. Thus a 'good sense of timing' not only indicates that we have a kairotic sense of when to take a specific action (Hutchings 2008) but also that we occupy a qualitatively intelligible and perhaps predictable realm in which it is feasible to do so, as in chronos. Likewise, a breakdown in timing or 'bad sense of timing' robs us of the capacity for effective intervention but also destabilises our larger worldview and sense that things hang together in an orderly, manageable fashion. Two implications of this view of timing are worth highlighting here. First, timing is always relative and positional, insofar as the relevance of various changes and the timing standard chosen are intrinsically matters of perspective. Second, and flowing from the first, timing is always for someone and something—it has a positionally-specific purpose.[4]

How do such timing activities and experiences produce the various 'times' or 'temporalities' we often find in IR, including linear, cyclical, and national-state variants (e.g. Walker 1993; Edkins 2003:1–13)? The 'figural' qualities of symbolic language, which tend to work 'in terms of reifying substantives' rather than dynamic relations (Elias 1989a:193–96, 1989b:342), provide the key link. For example, we regularly refer to the 'wind blowing' or a 'river flowing' even though wind is inseparable from its constitutive quality of blowing and a river is characterised by flowing (Elias 2007:36). Much the same, symbolic language transfigures the constitutive or characteristic features of a given timing activity into the objective or 'thingy' noun 'time'. So

---

came to feature more and more in symbolic systems. This was abetted by the figural qualities of human language (more on this below) (Elias 2007:43, 1989b:342). In other words, human capacities for symbolic language transposed features of timing activities into attributes of the noun 'time'. Over centuries these transpositions settled into a multifaceted conception of 'time'.

[3]   Although he does not spend any reflection on the difference, Elias (2007:39) suggests as much when he discusses the use of a human lifespan—which is non-repeatable and unpredictable—as a timing standard.

[4]   This connects timing with the critical IR of Robert Cox (1981).

when we hear about linear time in IR, an Eliasian perspective suggests that there is a timing activity behind it that synthesises changes as unique, sequential, and unidirectional; whereas a cyclical time refers to changes organised in a recurrent sequence that implicitly allows for progress or regress (as in 'cycles of violence'); while national-state time refers to the admixture of a particular origin story, national holidays, and the teleological promise of political perfection sedimented over time by successive iterations of collective identity.[5]

At the limit, our highly successful and widely institutionalised practices for timing a huge variety of activities using solar (clock) and lunar (calendar) cycles has, through symbolic language, produced the idea of a freestanding, existential dimension of 'natural time' (Elias 2007:96). Such linguistic reification serves the crucial purpose of transferring knowledge about a timing mode from generation to generation but it can also elide the dynamic and practical aspects of timing by convincing us that there are 'times' independent of human effort. If we want to better understand 'time', 'temporality', and their relationship to various socio-political processes, we need to foreground their dynamic and human origin in timing (Elias 2007:35) so that we do not unintentionally reify the very activities we are trying to unpack and in so doing elide the relative and positional character of any particular time or temporality, which only emerges to the extent that it ably serves some purpose for someone.

Identity as a Timing Standard

Identity formation is one such process. It has already been noted that identity formation is a dynamic and social practice with temporal qualities (Berenskoetter 2012; Prozorov 2011). We take no issue with such observations, but think that the discussion can be pushed much further by scrutinising the constitution of selfhood as a timing process with identity commitments as its timing standard.

In any timing activity, changes are synthesised into a coherent and orderly whole according to the timing standard. Although the aforementioned clock-based time reckoning is by now second nature for most, when timing in this way we implicitly prioritise the mensurable, quantified, and regularised changes produced by an internal escapement or quartz crystal (Landes 2000:8–18) as the frame of reference by which to coordinate our actions and

---

[5]   In IR, 'the time of the nation-state' has often been used less precisely to refer to some nebulous but generally unilinear-progressive promise thought to adhere to the sovereign statist way of doing politics (see Shapiro 2000; Edkins 2003:xiv; Stephens 2010:34; Lundborg 2011).

lives. 'Let's meet at 9:30' has no functional meaning if we do not both possess an understanding of this way of timing—that the '9' refers to a morning hour of day and the '30' to minutes, or half an hour beyond—as well as clocks coordinated to the same time zone, which allow us to plan ahead in order to arrive at the café at the same time.[6]

Clock reckoning is the most prevalent and likely intuitive example of a timing standard today, but it is in no way exclusive. Narrative theorists like Paul Ricoeur (1984, 1985, 1988) and David Carr (1986) have shown how the theme of a story provides a synoptic vision by which elements of experience are chosen, combined, and ordered to produce a followable sequence of events or vision of time leading as if by necessity to a conclusion that marks the narrative's chronological endpoint but also its hermeneutical fulfilment (see Hom 2013 chp. 2). In narrative, the theme provides the standard by which the story's other components are selected, integrated, coordinated— that is, by which the story is timed into a workable 'plot' that unfolds a particular temporality.

Given how important narrative is to a sense of self (Steele 2008; also Freeman 1993) we suggest that all this points to the idea that identity, or 'who' the social agent thinks she is, provides the timing standard by which to construct her self. In most autobiographies, the chronological conclusion also renders the subject in full. Inasmuch as the theme of a story is a timing standard and the theme of any autobiography is the self, we can think of identity commitments as the timing standard for the story of the self.

Along these lines, Solomon's (2014:679) recent work on time and subjectivity shows social identities are not fixed, stable, and comprehensive attributes but rather fluid and 'temporally decentred.' This means they necessarily contain forward- and backward-facing perspectives that make sense of past experiences and future possibilities in a way that allows the subject to come into being by acting in the present. This idea of temporal decentring calls out the importance of timing to identity construction.

Identities are constructed through retrospective autobiographical or 'self-reflexive' monitoring of how various past events experienced and actions already taken 'fit' into one's story of her self, where 'fit' refers to how consistent those actions and experiences are with her overarching idea of who she is (see Steele 2008:60–63). The overarching identity a social agent adopts also exerts a prospective or futural pull insofar as she envisions 'how someone like me would act if/when certain situations arise'. For Solomon (2014: 672), drawing upon Lacanian psychoanalytic theory, the subject is

---

6   Elsewhere, Hom (2010, 2012) has dubbed this 'Western standard time'.

continually constructed through a retroactive temporality, where the subject posits itself as 'having always been' a full or 'whole' self. Yet, this is a kind of 'fantasy' (in Lacanian terms), since the subject is always marked by a lack, a perpetual sense of incompleteness. Subjects are never fully self-present, since there is no extra-discursive or biological foundation upon which subjectivity ultimately rests. The subject is unstable, in this sense, insofar as it can never hook into a fixed foundation once and for all. This instability of subjectivity is part and parcel of the subject 'not having fully been' in the past yet 'not quite yet being' in the future. The subject only ever 'will have been' since it never fully reaches the future image of wholeness and centredness that it strives towards in its identity practices. As Lacan (2006: 78) puts it, the process of subject formation 'is a drama whose internal pressure pushes precipitously from insufficiency to anticipation.' The subject is thus an open self, produced at the ever-changing nexus of these 'backward' and 'forward' temporalities and constituted through identity practices simultaneously facing in both temporal directions (Solomon 2014: 674). Both retro- and pro-spection come together in an evolving present, where the subject takes actions or reacts to situations in accordance with her commitments. All of which is to say that the social agent's identity poses the standard by which all the changes she experiences are integrated and coordinated into a coherent, consistent, and intelligible story of her life that produces and confirms who she is.[7]

By helping us determine what matters, how to (re)act, and when to do so, identity provides a timing standard. When successful, identity commitments help a social agent to produce a smooth and compelling account of herself that unfolds in an orderly and predictable manner—e.g. from birth through schooling to a successful professional career. When unsuccessful, the social agent's life may seem disjointed, unpredictable, and devoid of intelligible connections between her past development, her present state, and her future direction. All of these latter outcomes suggest some misfit, either between her chosen identity and her environment (e.g. a very slow runner who identifies as a sprinter) or her ability to integrate and coordinate herself to that environment (e.g. 'I should have known better' or 'I knew what to do but I just couldn't bring myself to follow through') and will thus require much more work to transform into a smooth and intelligible story of who she is. Most importantly, perhaps, unlike a wholly retrospective story, in lived experience the adequacy of one's chosen identity as a standard for timing her life is an open practical question, just as the stability of her sense of self is an ongoing and constant task of simultaneous retroactive and prospective (re-)

---

[7]   While we do not argue that this is the *only* temporal account of identity formation, we do suggest that it may travel beyond the particular issues that we examine here. Many disciplines, for example, have productively drawn upon Lacan's theories of subject formation, including geography (Kingsbury 2007), urban planning (Gunder 2005), and organisation studies (Contu, Driver, and Jones 2010).

production.

These backward and forward-facing orientations of the temporally decentred self also mimic the way in which timing enables the serial perception of time as such. Although we tend to think of generally linear time as something external to ourselves along or through which we move, this is a product of timing. Creative synthesis enables this when we recall 'distinctly what happened earlier and … seeing it in [our] minds' eyes as a single picture, together with what happened later and what is happening now' (Elias 2007:31). In the case of serial time consciousness, this is a most basic sort of timing that coordinates the changing mind with a continuum of external stimuli and facilitates 'the perception of events which happen one after another as a "sequence in time"' (Elias 2007:31). When it concerns identity construction, we are usually interested in a more selective yet lengthier continuum of changes pertinent to who we think we are. Rather than the welter of immediate and often mundane experience from which we select and arrange elements such that 'time' seems to 'pass' one moment after another, our identity is based on more important and older recollections that we configure into a coherent narrative from which our sense of self springs.

The forward-facing temporality of identity similarly mirrors our basic prospective grasp of the continued flow of time (see Husserl 1964; Ricoeur 1984; Carr 1986). In basic time consciousness, we intuitively grasp what is likely to happen next based on our immediate context—e.g. if a ball is rolling along a table we expect it to fall towards the ground once it passes the table edge. If something novel happens instead—e.g. if the ball disappears or falls upwards—our 'sense of time' may falter.[8] If only one unexpected thing happens we may perceive that 'time stands still', but if many unexpected things happen at once or in quick succession we may perceive just the opposite, that 'time moves too quickly' or that 'chaos' replaces any apprehensible temporal flow. Once again, these possibilities and their implications for time are intensely relative and positional—that is, they pertain to our sense of time or feeling of chaos, which may not seem as such to others.[9] Much the same, our sense of who we are entails expectations about who we will become[10] and how we should and will react to future

---

[8]   Although the example posed here is utterly banal, this relationship between novel change and time consciousness lies at the heart of trauma research, which argues that acts and experiences so unprecedented and violent as to be 'unspeakable' actually 'rupture' our sense of time because we cannot accommodate them in our ongoing perception of how things fit together (see, e.g. Edkins 2003).

[9]   For example, a surprise attack might 'sow chaos' through a military base and cause soldiers to feel that time is 'rushing by' but also represent the orderly unfolding of a temporality planned and executed by the sneak attackers.

[10]   For a related discussion, see Berenskoetter's (2011) investigation of the temporal

developments. If our expectations are confounded by events that are unprecedented or not easily accommodated in our autobiography, this may throw off our sense of time as well as challenge our identity because we do not know how to react, which in turn causes our well-ordered serial perception and our coherent, comprehensive, and clean vision of our self to falter.

These dynamics and tensions in identity construction and maintenance are well-known to social construction theorists, who often characterise self-identity as a sense of personal continuity or 'ontological security' that persists 'through' or 'over' time (Steele 2008; also Freeman 1993). However, in the account proposed here, both identity and time are matters of timing insofar as they involve a thinking subject who integrates and coordinates her personal change continuum with a variety of experiences in order to preserve or reproduce both a sense of orderly temporal continuity and a sense of coherent self. On this view, instead of understanding self-identity as a matter of establishing continuity 'through' or 'over time', it is more useful and accurate to approach it as a matter of self-identifying through timing. Similarly, the way in which a subject is 'stitching across' (Solomon 2014:675) multiple temporalities and constantly negotiating changes in the past and future that appear external to it underscores that subjectivity is an active, ongoing, and continuous effort to synthesise an autobiographical vision with multiple changes in the world of experience in a way which reaffirms one's self as consistent, 'full', and stable. In other words, it is an individual timing process that requires us to encompass our past and future in order to enable both action in an intelligible 'present' and an overarching, consistent, and coherent idea of who we are.

Timing, Identity, and Desire

It is through the nexus of retroactive and anticipatory temporalities that the emotional element of desire arises.[11] Following Lacan (2006), we understand desire as the continual longing for ontological stability and wholeness which, though ultimately illusory, the subject nevertheless continually seeks.[12] Desire emerges through the subject's 'lack' – that it is never fully present in a discursive or temporal sense. This lack of—and desire for—a 'whole' identity is what sparks the subject's identity practices, even in the face of continual

---

qualities of visions.

[11]   While IR has recently shown much interest in affect and emotions (Crawford 2000; Fierke 2013; Hutchison and Bleiker 2014; Ross 2014; Solomon 2012), little of this work has explored the relationships between time, timing, and identity.

[12]   For a more detailed discussion of the relationship between desire, discourse, and subjectivity, see Solomon (2015).

frustration in never quite achieving the pursued sense of self-unity (Solomon 2014: 675). This is why Lacan terms such identity narratives as 'fantasies', in the sense that while they may seem to subjects to promise ontological stability, they can never fully deliver this since lack continually re-emerges due to the subject's absence of foundations. Since desire is an effect of lack, it can never be fully satisfied or extinguished. As such, the subject is always a subject of desire; desire is the propellant pushing the subject to continue its identity narrating practices.

Slavoj Žižek offers a political example of how these temporal dynamics help to elicit subjects' desire to reach stability via narratives of the nation-state. He (2002: 197) argues that the 'basic paradox of the psychoanalytic notion of fantasy consists in a kind of time loop—the 'original' fantasy' is always the fantasy of the origins.' Here, one of the most common political narratives is the foundation myth of the nation. A nation's politics often revolve around efforts to re-narrate the nation's 'origins' in order to co-opt them into contemporary agendas. Yet as Žižek (1993: 127) points out, national origins 'are never simple given facts: We can never refer to them as a found condition, context, or pre-supposition of our activity. Precisely as presuppositions, such narratives are always already posited' by us. Tradition is tradition insofar as we constitute it as such' (Žižek 1993:127). The national desire for ontological stability via collective unity leads to the positing of a 'fullness' that never in fact existed at the supposed point of origin (Solomon 2014: 675).

These insights can be extended to more fully draw out how the entanglements of desire, temporality, and identity often function at collective levels. For example, the social construction of the post-9/11 war on terror can be more comprehensively analysed through a temporal-desire lens. Many constructivist studies of the war on terror have emphasised longstanding themes of identity in IR, such as the argument that the war on terror was a historically contingent discourse rather than a natural one, and that the language of 'evil,' 'civilization,' and 'barbarism' produced a series of identities that were relationally structured through binaries of self/other (see Croft 2006; Holland 2009; Hülsse and Spencer 2008; Jackson 2005). Yet, it is the interweaving of timing and desire that can provide a more comprehensive picture of the social construction of identity in the war on terror.

The social construction of the subject 'America' in the war on terror points to not only the contingency and othering processes involved. It also reveals the role of this retroactive temporality and desire that helped to pull subjects to identify with its narrative. In a 20 September 2001 speech before a special joint session of the U.S. Congress, President George W. Bush offers a narrative of the events nine days before. Here, the collective subject of the nation' ('America') suffered a 'wound'. Three thousand lives lost is recognised as a tragedy, yet throughout the discourse there seems to be something else that was lost on 9/11. Bush alludes to it several times and in several different ways, something beyond the loss of lives: 'Great harm has been done to us. We have suffered a great loss'; 'night fell on a different world, a world where freedom itself is under attack'; 'a threat to our way of life.' 'These terrorists kill not merely to end lives, but to disrupt and end a way of life.' 'Terrorists attacked a symbol of American prosperity. They did not touch its source. America is successful because of the hard work, and creativity, and enterprise of our people.' 'This is a fight for all who believe in progress and pluralism, tolerance and freedom.' (Bush 2001) These discursive attempts to pin down exactly what was lost – yet without ever definitively doing so – illustrate the prospective fullness that is seen as desirable. 'Our way of life' and so on is discursively as close as one can get to the perceived 'essence' of the nation. Yet, these are ways of covering over the incompleteness—the lack—of a 'whole' nation by re-synthesising, that is, re-timing 'America' as a coherent and complete entity intelligibly oriented and capable of effective action in a ruptured world. Bush's numerous attempts to discursively 'button down' what exactly 'we' are points to the very indefinability of that which must be defended from the terrorist enemy (Solomon 2014: 677-78; Solomon 2015). The war on terror narrative (re) constructs the fantasy of a 'whole' 'America' (the subject of the discourse) that is partially represented in discourse, yet is also 'missing' something that is posited as central to its 'sense of self'.

The ideal of a 'whole' nation free of threats and ambiguities is an image that covers over the constitutive lack of such an entity. A unified 'America' is posited as lost, yet, such an 'America' did not, in fact, exist before 9/11. This retroactively projected ideal of 'America' is posited as having been lost at the moment of trauma, yet it had never been fully constructed before then. It must be assumed to have existed, though, in order for the war on terror discourse to be meaningful. In this sense, the retroactive presupposition of the ideal drives the desire for it, and helps to account for the 'pull' factor that elicited audiences' identifications with the war on terror discourse in the wake of 9/11 (Solomon 2014: 679).

Assessing identity as a timing project locates the driver of desire as part of the process of the ongoing drama of the subject, i.e. the timing agent. His desire for fullness is precisely a desire to comprehensively coordinate his own personal change continuum with those changes that impact him and thereby gain a whole 'sense of self' and a 'sense of time'. We might say here that a sense of self and sense of time are co-constitutive products of effective timing. On this view, desire cannot be reduced to some primitive id or biological drive, nor does it depend on some subject-independent and antagonistic past and future (see Solomon 2014:675). Instead, desire wells up from the challenges of internal, subjective, and identity-based timing, a difficult and continuous work that, when less successful, leaves us feeling incomplete, 'partly present' yet 'partly missing' (Solomon 2014:678) or, when more successful, grants a sense of closure-in-wholeness, of consistency with the world around us, and of 'being present'. As Johnston (2008: 260-1) argues, desire arises from the subject's construction through these overlapping temporalities. 'Drives and desires aren't simply quotas of energetic force welling up from the brute corporality of a primitive id. Rather, they are the aftershocks generated by the repeated collisions of incompatible temporal dynamics' (Johnston 2008: 260-1). In this sense, the subject's attempts to stabilise its self-narrations is a form of timing in its own right. Crucially, such fullness or completion is always temporary. Good timing never lasts and we are never fully timed due to uncertainty, ambiguity, complexity, and emergent changes that require further synthesis—while we desire closure, we remain engaged in an open-ended project to time our self to the world disclosed to us through experience.

In these ways, a theory of timing helps locate emotion and desire within the dynamic construction of the self. Emotion and desire reflect back on timing, helping to further clarify the stakes and challenges of timing as a basic human activity. Because timing is never finished so long as meanings remain fluid or change continues to occur, the issues that impel timing in the first place are ever-present, namely: complexity, puzzles about how things hang together and how we relate to them, concerns about novel changes, and a variety of questions about who we are. Successful timing reduces complexity, proffers a working model of how things hang together and where we fit in, accommodates novel changes in a smooth trajectory, and helps establish our sense of self. However, if any of these issues present intractable difficulties— if complexity shades into utter chaos, if we cannot explain a surprising development, or if we cannot establish clear links between our sense of self and our environment—timing falters and with it our sense of wholeness, so we must re-interpret or 'retro-fit' the problematic experience, revise the timing mode, or search for a new standard altogether.[13] As may be apparent, these possibilities entail significant effort and—as we move closer to jettisoning our timing standard in search of an alternative—pose the limit possibility of an utter breakdown in timing: the possibility that no standards or modes of synthesis can be found that are adequate to all the pertinent aspects of a developing situation. In such a setting, our sense of time as well as any confidence about who we are comes under severe threat.

---

[13] See Hom (2016) for further discussion and an example of retrofitting drawn from scholarship of the 'Arab Spring'.

These intimate links between a solid sense of self and a stable sense of time highlight the emotional and desirous nature of timing in general. It makes sense that identity commitments produce a desire for stability and fullness in the sense of a consistent, synoptic, and coherent sense of self. However, we think this sort of desire can be scaled 'up' from the individual's conception of her own self to her model of how the wider world in which she finds herself hangs together. After all, it is not just that the self must remain stable and coherent 'through time'; time as such must also seem to flow in a manageable way or else it threatens to overwhelm the self with unremitting flux, disorder, and chaos.[14] Put differently, it is not just that we want a stable sense of who we are, we also desire a stable and full sense that the time we inhabit is consistent, reliable, and orderly rather than some onrushing and untamable river of unintelligible events. Thus, just as acts or experiences that confound our sense of self may engender identity crises replete with emotional facets, unexpected developments may signal breaks or ruptures in the smooth flow of time that engender fear, anxiety, and dread in us. This is because both augur an utter dissolution of timing, the thoroughgoing failure of a chosen standard to provide orientation, intelligibility, and control by synthesising relevant environmental changes with our own personal continuum. Such a problem can only be resolved by colossal efforts and may require us to completely change our understanding of how the world works and/or who we actually are. No wonder, then, that we desire a fullness of self in a manageable flow of time, for both are necessary conditions for a social agent to go on in life.

This framework can be scaled 'up' in a second sense, from the individual agent to the sort of collective selves from which international politics is composed. The earlier discussion of 9/11 shows how a state works to reconstitute its sense of self and of time after a shocking experience. Indeed, if anything, timing of a collective self is much more difficult and politically fraught than timing an individual identity. The timing standard of identity commitments must synthesise a much larger quantity and range of change continua, the potential sources and sites of novel changes that disrupt a collective self and its sense of time are far greater,[15] and the backward and forward temporal views often extend much further than in an individual life since the nation is both older and more durable than its constituent citizens (see Niebuhr 2001). All of these features proliferate the complexity of the timing challenge involved in identity formation and maintenance. They also up its stakes inasmuch as both political practitioners and subjects, on the one hand, and the international system itself, on the other, depend on raison

---

[14] These problematic qualities are some of the oldest attributed to time's natural essence or force (see Brandon 1965; Hom 2012).

[15] This includes 'internal' sources in the form of domestic actions or challenges to the state's authority (see Steele 2010:133–64).

d'état for their raison d'être.

Conclusion

In making the case for the emotional/desirous aspect of timing, this chapter hopes to raise a few key issues for the study of time and identity in IR. First, despite the recent interest in time and temporality in IR, the notion of timing can be viewed as the more primary process through which various conceptions of 'time' are produced. The use of language contracts our notions of time and transfigures the constitutive aspects of a particular timing activity into the noun 'time'. This holds insight for understanding processes of identity and subject formation. While IR scholars have long interrogated the social construction of identity, and although others have recently pointed out the importance of time to identity, we suggest that timing may prove to be a more fruitful avenue through which to investigate the temporal politics of identity. Viewing identity itself as a timing process, we argue that identity poses the standard by which changes the subject experiences are integrated and synthesised into a coherent and intelligible self-narrative. Identity, then, is not merely an image of a self in relation to an other (Wendt 1999). Rather, it is able to function in such a manner because of its prior function as a timing standard by which the self initially coheres.

Second, the chapter draws together this understanding of time and timing with the emotional element of desire. Although IR has recently shown much interest in emotions, we contend that examining the relationships among timing, identity, and desire offers a richer account of the social construction of subjectivity than most extant accounts. Drawing upon insights from psychoanalytic theory, the chapter argues that not only is timing a key element of identity, but also that this timing process occurs alongside (and likely because of) multiple overlapping temporalities. In this view, the subject's decentring is partly due to its construction at the nexus of these retroactive and anticipatory temporalities. It is this decentring – along with the subject's ontological lack – that sparks the emotional pull of desire. Subjects desire identity narratives that promise a closed timing mode in the form of a 'whole' or 'complete' identity, even if such wholeness is ultimately illusory. In short, the subject's efforts at stabilising its self-narrative in the face of continual decentring at the point of multiple overlapping and conflicting temporalities is a crucial form of timing. Timing, time, and emotion/desire, in this sense, compound a much more complex relationship than is normally recognised. This relationship is politically challenging because it is intrinsically positional, holds implications for others (either within or without this identity), and is contestable by them.

Finally, this understanding of the entanglement of timing, identity, and desire suggest a different strategy of critique of dominant discourses than IR usually recognises. Much of critical IR's understanding of critique lies in the general notion of de-naturalisation – that is, that effective critique must aim not at engaging a problem or issue necessarily on its own terms, but rather to show the historical and contingent constitution of the problem itself. In other words, critical IR often aims (rightly so) at demonstrating the contingency of contemporary arrangements – that since things could have been different, they can change and be changed. Cox's (1981) distinction between problem-solving and critical theory comes to mind here. Yet, as sensible as this seems, critiques of dominant discourses often fall flat. This chapter suggests instead that criticism should incorporate the temporal and desirous dimensions of identities. Criticism should be aimed not merely at pointing out the historical contingency of an issue. Rather, it should also aim at displacing the emotional investments in the identity narratives in question from wholeness and closure to an explicitly open politics of identity and timing. If social subjects can only ever engage in the open-ended timing of the Self, then we should explicitly develop a politics from this rather than pining for its impossible alternative (see Hom and Steele 2010; Hom 2016). By disrupting the timing standards and concomitant modes of coordination and control that implicitly underpin dominant identity narratives for the positive purpose of fostering a politics of open timing, and also thereby helping to shake up the investments of desire that fortify their hegemony, social criticism may be more able to make some headway against discourses of violence and control, such as the war on terror. If we enrich our models of subjectivity and identity by integrating notions of timing and emotion/desire, our critiques may be more effective in shaking these very elements that help to constitute the power and pull of identity.

References

Berenskoetter, Felix. (2011) Reclaiming the Vision Thing: Constructivists as Students of the Future. *International Studies Quarterly* 55: 647–668.

Brandon, S.G.F. (1965) *History, Time, and Deity*. Manchester: Manchester University Press.

Bush, George W. (2001) Address to a Joint Session of Congress, September 20. Full text at http://www.washingtonpost.com/wpsrv/nation/specials/attacked/transcripts/bushaddress_092001.html.

Callahan, William A. (2006) War, Shame, and Time: Pastoral Governance and National Identity in England and America. *International Studies Quarterly* 50: 395–419.

Campbell, Joseph Keim, Michael O'Rourke, and Harry S. Silverstein, Eds. (2010) *Time and Identity*. Cambridge, MA: MIT Press.

Carr, David. (1986) *Time, Narrative, and History*. Bloomington, IN: Indiana University Press.

Croft, Stuart. (2006) *Culture, Crisis, and America's War on Terror*. Cambridge: Cambridge University Press.

Cox, Robert. (1981) Social Forces, States and World Orders: Beyond International Relations Theory. *Millennium: Journal of International Studies* 10: 126–55.

Crawford, Neta. (2000) The Passion of World Politics: Propositions on Emotions and Emotional Relationships. *International Security* 24(4): 116-156.

Edkins, Jenny. (2003) *Trauma and the Memory of Politics*. Cambridge: Cambridge University Press.

Elias, Norbert. (2007) *An Essay on Time*. Dublin: University College Dublin Press.

Elias, Norbert. (1989a) The Symbol Theory: An Introduction, Part One. *Theory, Culture and Society* 6: 169–217.

Elias, Norbert. (1989b) The Symbol Theory: Part Two. *Theory, Culture and Society* 6: 339–383.

Fierke K.M. (2013) *Political Self-Sacrifice: agency, body and emotion in International Relations*. Cambridge: Cambridge University Press.

Freeman, Mark. (1993) *Rewriting the Self: History, Memory, Narrative*. Abingdon: Routledge.

Hall, Rodney Bruce. (1999) *National Collective Identity: Social Constructs and International Systems*. New York: Columbia University Press.

Holland, Jack. (2009) From September 11th, 2001 to 9-11: From Void to Crisis. *International Political Sociology* 3(3): 275-292.

Hom, Andrew R. (2016) Angst Springs Eternal: Dangerous Times and the Dangers of Timing the "Arab Spring." *Security Dialogue* 47(2) 165–183.

Hom, Andrew R. (2010) Hegemonic Metronome: The Ascendancy of Western Standard Time. *Review of International Studies* 36: 1145–1170.

Hom, Andrew R. (2013) Reckoning Ruin: International Relations Theorising and the Problem of Time. Ph.D. Thesis, Aberystwyth University.

Hom, Andrew R. (2012) Two Regimes of Time. *E-International Relations*. Available at: http://www.e-ir.info/2012/12/24/two-regimes-of-time/. (Accessed July 9, 2015).

Hom, Andrew R., and Brent J. Steele. (2010) Open Horizons: The Temporal Visions of Reflexive Realism. *International Studies Review* 12: 271–300.

Hülsse, Rainer and Alexander Spencer. (2008) The Metaphor of Terror: Terrorism Studies and the Constructivist Turn. *Security Dialogue* 39(6): 571-592.

Husserl, Edmund. (1964) *Phenomenology of Internal Time Consciousness*. Bloomington, IN: Indiana University Press.

Hutchings, Kimberly. (2008) *Time and World Politics: Thinking the Present*. Manchester: Manchester University Press.

Hutchison Emma and Bleiker Roland. (2014) Theorizing emotions in world politics. *International Theory* 6(3): 491-514.

Jackson, Richard. (2005) *Writing the War on Terrorism: Language, Politics, and Counter-terrorism*. Manchester and New York: Manchester University Press.

Landes, David S. (2000) *Revolution in Time: Clocks and the Making of the Modern World*. Cambridge, MA: Belknap Press of Harvard University Press.

Lapid, Yosef, and Friedrich Kratochwil. (1996) *The Return of Culture and Identity in International Relations Theory*. Boulder, CO: Lynne Rienner.

Lundborg, Tom. (2011) *Politics of the Event: Time, Movement, Becoming.* Abingdon: Routledge.

McIntosh, Christopher. (2015) Theory across Time: The Privileging of Time-Less Theory in International Relations. *International Theory* 7: 464–500.

Niebuhr, Reinhold. (2001) *Moral Man and Immoral Society: A Study in Ethics and Politics.* Louisville, KY: Westminster John Knox Press.

Ricoeur, Paul. (1984) *Time and Narrative, Volume 1.* edited by David Pellauer. Chicago: The University of Chicago Press.

Ricoeur, Paul. (1985) *Time and Narrative, Volume 2.* edited by David Pellauer. Chicago: The University of Chicago Press.

Ricoeur, Paul. (1988) *Time and Narrative, Volume 3.* edited by David Pellauer. Chicago: The University of Chicago Press.

Ringmar, Erik. (1996) *Identity, Interest and Action: A Cultural Explanation of Sweden's Intervention in the Thirty Years War.* Cambridge: Cambridge University Press.

Ross, Andrew A. G. (2013) *Mixed Emotions: Beyond Fear and Hatred in International Conflict.* Chicago and London: University of Chicago Press.

Shapiro, Michael J. (2000) National Times and Other Times: Re-Thinking Citizenship. *Cultural Studies* 14: 79–98.

Solomon, Ty (2012) 'I wasn't angry, because I couldn't believe it was happening': Affect and Discourse in Responses to 9/11. *Review of International Studies* 38(4): 907-928.

Solomon, Ty. (2014) Time and Subjectivity in World Politics. *International Studies Quarterly* 58: 671–681.

Solomon, Ty (2015) *The Politics of Subjectivity in American Foreign Policy Discourses.* Ann Arbor: University of Michigan Press.

Steele, Brent J. (2010) *Defacing Power: The Aesthetics of Insecurity in Global Politics.* Ann Arbor, MI: University of Michigan Press.

Steele, Brent J. (2008) *Ontological Security in International Relations: Self-Identity and the IR State*. Abingdon: Routledge.

Stephens, Angharad Closs. (2010) Citizenship without Community: Time, Design and the City. *Citizenship Studies* 14: 31–46.

Stockdale, Liam P.D. (2013) Imagined Futures and Exceptional Presents: A Conceptual Critique of "pre-Emptive Security." *Global Change, Peace & Security* 25: 141–157.

Walker, R. B. J. (1993) *Inside/Outside: International Relations as Political Theory*. Cambridge: Cambridge University Press.

Wendt, Alexander (1999) *Social Theory of International Politics*. Cambridge: Cambridge University Press.

Žižek, Slavoj (2002) *For They Know Not What They Do: Enjoyment as a Political Factor*. Second Edition. London and New York: Verso

Žižek, Slavoj (1993) *Tarrying with the Negative: Kant, Hegel, and the Critique of Ideology*. Durham: Duke University Press.

2

Time Creators and Time Creatures in the Ethics of World Politics

KIMBERLY HUTCHINGS
QUEEN MARY UNIVERSITY OF LONDON, UK

Truth as the Father of Time

Rousseau's jibe, that Hobbes showed us not man but Englishmen in a state of nature, makes a profound and perennially relevant point about the nature of rationalist theorising about ethics and politics. Specifically here my concern is with ethical theorists who believe that answers to ethical questions about world politics can be arrived at through the exercise of impartial reason at the level of ideal theory, often using thought experiments as a way of clarifying and resolving moral dilemmas about war or global justice (see Hutchings 2010: 28-53 for an overview of rationalist ethical theories; Simmons 2010 for a discussion of the ideal/ non-ideal theory distinction). Thought experiments construct scenarios in order to test out moral intuitions, and may range from Singer's famous use of the example of the passing adult's obligation to save a child from drowning in order to demonstrate the moral obligations of the globally affluent towards the starving (Singer 1972) to increasingly fantastical articulations of the 'trolley problem' to tease out circumstances in which killing the innocent may be justified.[16] This kind of moral reasoning is particularly characteristic of certain trends in cosmopolitan moral theory based on deontological, consequentialist or contractualist assumptions, especially recent developments in the ethics of war (see McMahan 2009; Fabre 2012).

As critics of ideal theory have persistently pointed out, the principles, actors

---

[16] see: https://en.wikipedia.org/wiki/Trolley_problem

and situations that make up the worlds of thought experiments tend to reflect and incorporate assumptions that are specific to the time and place of the theorist, whether the theorist is conscious of this or not (Mills 2005; Miller 2008). In what follows I suggest that one such set of assumptions that remains in place, and continues to do a lot of work in the articulation and reception of rationalist international ethical and political theory, is a set of assumptions about the temporality of world politics. The ways in which traditions of moral thought are mobilised in ethical rationalism is indifferent to historical context. But this very indifference is premised on assumptions about time and the role of the theorist in relation to it that are highly historically specific. They are assumptions that follow from the subsumption of moral theory under an epistemic, Baconian model of science (Hutchings 2008: 32-34). On this account, truth (science) has priority over time (nature) and provides the key to controlling and making time (second nature). Within this worldview, as I have explained elsewhere, the temporality of the ethical and political present (within the possible worlds of ethical argumentation and thought experiment) is a temporality of creative action in which creative time-making (*Kairos*) imposes itself on natural, chronological time (*Chronos*). In keeping with Baconian rationalism, agents within the simplified worlds of ethical theory, who are depicted as making decisions about what it would be right to do in particular situations (e.g. when faced with choices between killing and letting die in the contest of war) are, *insofar as they enact true principles*, time-makers. This temporal positioning is confirmed by the fact that the imagined situations in which agents in these scenarios are caught, necessitate that the capacity to master time is not universally shared. Shadowy 'other' actors accompany the moral agents in the spotlight of the possible world. Without these other actors, there is no situation, there are no moral dilemmas and questions.

The focus in scenarios imagined by ethical theorists is on the question of what would be right in certain situations for moral agents to do, but in order for this question to be raised moral agents require the presence of those who are not competent moral agents in the fullest sense. They are the mistaken, the wicked, the ignorant and the incapable. Without them, no moral dilemmas would arise. The temporality of the simplified world of ethical thought experiments, therefore, is double and paradoxical, it is one in which time creators and time creatures share and do not share the same temporality of action, the (hypothetical) present. This unevenness is underlined by the abstracted practices and institutions that also shape the situation, and that reflect a skew towards the perspective of those moral agents for whom certain questions are particularly relevant. Questions about the criteria for intervention, about who should be killable and who should not, about whether there is an obligation to give aid, to reform global governance, or to redistribute wealth. Such questions presuppose asymmetrical power relations within the simplified world and are addressed to the strong, who are strong in two interrelated senses: first in the sense that they are full moral agents; second, in the sense that they occupy positions of power.

The scenarios constructed by ethical theorists do their work at the level of ideal theory through the identification of the theorist (truth teller) and their audience with idealised moral agency. The purpose of the exercise is to illuminate what should be done in a context in which what ought to be done can be done. This means that the theorist and audience see their situation from the perspective of the time-makers. This identification is cemented through the examples and analogies used to bring the argument home. Where these illustrations are fictional, they typically place the moral agent in a relation in which moral hierarchy and asymmetrical power are obvious: adults and children; perpetrators and victims; punishers and criminals. Where these illustrations are historical, they typically draw on events specific to histories and experiences that will be familiar to theorist and audience. The reason, I suggest, why these kinds of illustrations are so important for rationalist moral and political theory is that they make clear the homology between the fictional temporality of the possible world of thought experimentation and unquestioned assumptions about the temporality of actual world politics, and thus enable theorist and audience to recognise themselves within the imagined scenarios through which they develop their arguments. These unquestioned assumptions not only presume a world in which the time of action is unevenly distributed, but map this uneven distribution so as to identify the truth that generates time with the time and place of the theorist and audience. Possible worlds incorporate precisely the assumptions of simultaneous equality and hierarchy that characterise predominant narratives of world politics and economics in the post-Cold War period. Moreover, they reproduce the same mapping of strength and weakness, goodness and badness and the same recipe for how weakness and badness may be addressed through the good offices of the strong and the good, the time makers amongst us. In effect, the possible world accomplishes for moral rationalism, what philosophical history accomplishes for historicism; it grants a universal and unifying significance to the time making powers of western modernity for the world as a whole.

The implications of the reproduction of the western moral imaginary for the practice of theory are evident in Singer's previously cited argument for the moral requirement on individuals to give aid to famine victims. In order to make his point, Singer makes an analogy between the obligation for the affluent to give aid in a specified situation of famine in which certain conditions are assumed, and the obligation of an adult passing by to save a child from drowning in a puddle. Singer's adult/ child analogy works powerfully to dramatise his point about the nature of the obligation of affluent westerners to aid victims of famine. But it does so by explicitly characterising the relation between the affluent west and the victim other in terms of a protective relation between the mature adult and the immature child. Singer's argument could not be addressed to the victim/ child, because the latter is a subject position defined through incapacity, an incapacity that, by implication, is spiritual as well as material. What is being set up is not just a hierarchical way of seeing the world, but also the exclusion of victim/ child from the moral conversation, because in addition to being incapable of action, the victim/ child does not know anything morally or empirically valuable. This endows the ethical theorist and the addressees of his arguments with a position of ethical privilege, one that empowers both theorist and audience, within the moral imaginary of the simplified world, to act unilaterally because of their grasp of moral and empirical truth. Moral reasoning, on this account, becomes a profoundly hierarchical and exclusive business, even while it aims to forward universal principles of moral equality. The moral equality of theorist, audience and hypothetical moral agents in the possible world is inseparable from a distinction between those who grasp moral equality as a truth and those that do not.

To suggest that this is corrupting of the ambitions of rationalist cosmopolitan arguments may sound hyperbolic, but I do not think it overstates the case. It is corrupting because of its implicit identification of an audience capable of learning with those who already share the same world, and because of the limitations that it places on what can be learnt. And also because of the dispositional relations it sets up between moral theorists and those on behalf of whom they claim to speak - the shadowy, less than fully competent agents, who are stuck in time and not, by definition, going to learn from their role in the moral theorists' thought experiments. The explicit purpose of such arguments is to provide critical standards for thought, yet scenarios are constructed in such a way that no fundamental shifting of established ways of thinking about the ethics of world politics is possible. Thus one may argue over the principles or even the permissibility of aid, development, resistant violence or humanitarian intervention, but always within the terms of moral and empirical reference points that are fused within a particular moral imaginary. Those terms keep meaningful conversation to a restricted group of participants – perhaps best captured by the term 'the beneficent powerful'. To engage with the moral imaginaries of the vulnerable, the immature, the wicked, or the incapable would be to 'go back' to times that have been transcended by truth. Each time cosmopolitan ethical theorists replay their account of moral agency situated in the context of its shadowy others (perpetrators and victims), the moral superiority and historical advantage of the theorist and of their audience is reaffirmed.

The co-existence of equality and inequality in the worlds of moral rationalism is rendered comprehensible by being embedded in a temporal imaginary specific to Baconian assumptions about the relation between truth and time. Within this imaginary, the dispositional relation between the moral theorist and those on behalf of whom she speaks most commonly manifests itself in one of three modalities: *protective*; *educative* and *punitive*. Moral rationalists use their simplified worlds to enable identification between theorist, addressee and idealised moral agency. This means that they identify with the moral position, choices and dilemmas of the protector, the teacher and the law enforcer. This way of approaching ethical questions is epistemic and technocratic, for rationalist moral theorists the holder of moral truth within the possible world is beyond the reach of affect and power, and able to be effective in implementing the requirements of truth. In this respect, although theorists and their addressees make mistakes, these mistakes are of a particular kind, they do not disturb the fundamental subject position of the truth teller and truth seeker. Of course, theorists know that they can get things wrong – as is evident from ongoing disputes between different rationalist positions – but they can never be fundamentally wrong in the world-shifting way that time creatures are wrong. At the same time as having a conversation between themselves, therefore, a hierarchical relation with others is being reproduced. In this respect, the moral sensibilities and sensitivities cultivated in the hypothetical worlds of moral rationalism are those of authority derived from superiority.

Truth as the Daughter of Time

It might be argued that the corruption I identify follows not from the assumptions of moral rationalism but from the illicit smuggling in of elements of western bias into imagined worlds, for instance through the analogy between developed world and adult. If this is the case, then it implies that a moral technocracy that can demonstrate to all what ought to be done through access to the realm of truth (beyond time) is a viable possibility. But unless the actual world has already achieved moral perfection from the point of view of all (in which case the moral technocracy would be redundant) then this would still leave the hierarchy between time creators and time creatures, and the reproduction of that world in place. Although it undoubtedly provides a way of ensuring rigorous clarification of what the truth is claimed be, any normative theory modelled on Baconian assumptions elevates those that grasp the truth over those that don't. When it comes to normative theories about world politics it is hard to see how any particular band of moral technocrats can avoid reproducing the hierarchies that have enabled the leisure and provided the tools to exercise that technocracy in the first place. If we are to counter this tendency, then the chances are higher if rather than seeking to inhabit an imaginary that is completely out of this world, we replace Bacon's vision of truth as the masculine progenitor of time with the alternative conceit of a feminised truth, the daughter of time, and use this to multiply the worldly imaginaries that are expressed in our hypothetical worlds.

If moral truth is the daughter of time rather than its father, then there are no timeless ethical truths that have the capacity to make time, and the temporal parameters of judgment have to become an explicit focus of attention for international ethical and political theory. The coincidence of moral superiority and historical advantage that is rationalised in the possible worlds of moral rationalism (and the developmental philosophies of history of liberalism and Marxism) relies on accounts of world political temporality as singular, unevenly progressive and led by the West. If this temporal framing is put into question, then so is the mutual reinforcement between claims to moral superiority and historical privilege, and therefore between moral superiority and particular ways of organising human relations: interpersonally, socially, economically and politically. If one is interested in doing international ethical and political theory in a way that does not take the modernist moral/ historical link for granted, then provincialising work is necessary. This is essentially work of *disorientation* and *reorientation* as part of the construction of moral imaginaries less hidebound by the assumptions that structure the possible worlds of rationalism. So, how is this to be done? And what are the implications in terms of the concerns of international ethical and political theory?

First of all, the moral theorist has to recognise and acknowledge the moral imaginary that he or she takes for granted. Moral rationalism encourages the equation of a moral imaginary with a set of epistemic premises, but in practice no moral imaginary is confined to articulable ethical principles and values, it also includes assumptions about situations, about protagonists, about empirical facts, about lived experience and about the space and time of moral engagement. All of this, I suggest, when trying to think ethically about *world* politics is underwritten and made intelligible by a reading of world political time, a set of assumptions about the nature and meaning of the world political present in relation to past and future, and the place of one's theoretical voice within that narrative. The process of disorientation I am recommending requires the rationalist moral theorist to leave the comfort zone of this identification between moral superiority and their own particular place and time and to embrace an 'out of jointness' in which their own 'backwardness' or 'wickedness' could be possible. This may sound like a reintroduction of some kind of transcendental move beyond time, but this would only be the case if the temporality of ethical judgment prevalent within the western academy is the only possible temporality, and if truth is the daughter not the father of time then that is something that could not self-evidently be the case. In fact, we *know* it is not the case, in the banal sense that the moral superiority/ historical advantage story has not always framed ethical and political thinking and has been, and continues to be, consistently contested from inside and outside the political communities that have been its key proselytisers. The only way to suspend dependence on a particular reading of the present is to open up one's moral imaginary to other orientations. Very often, historically, such opening up has been violently enforced rather than willingly embraced. Nevertheless, for most of the populations that occupy the shadowy 'other' positions in the possible worlds of moral rationalism, engagement between different moral imaginaries and different temporal orderings is commonplace. Borrowing Chakrabarty's terminology, I have suggested elsewhere that international ethical and political theorists need to cultivate a heterotemporal orientation towards ethical judgment (Chakrabarty 2000; Hutchings 2011).

A heterotemporal orientation to cosmopolitanism decentres the position of the ethical theorist by questioning the assumption of a fusion between his or her particular present and 'the' present of world politics. It raises the question as to why, for example, humanitarian intervention or the 'Responsibility to Protect' should be taken as a sign of the distinctiveness of the world-political present. For whom, and from whose perspective is this a novel development? Does it mark a normative difference in the conduct of world politics or simply confirm a set of longstanding patterns? To raise the question of novelty is to disturb the kinds of subjective certainty, of 'at homeness' in thought, that render phenomena such as humanitarian intervention straightforwardly timely. In this respect, a heterotemporal orientation makes the work of the theorist much harder, since it requires the painful, political effort of cross-temporal engagement without the short cuts enabled by the taken for granted fusion of his or her particular present with the end of history.

If humanitarian intervention is identified with the potential globalisation of justice, then a heterotemporal orientation would suggest that what is needed is to begin by acknowledging and examining political temporalities of violation, in order to understand the meanings of injustice in the present. This would enable judgment of the likely effects of the institutionalisation of particular normative priorities in the principles and practices of international humanitarianism. But it would also open up the question of what kinds of violation matter and why, and offer a different route to the establishment of international hierarchies of outrage than that reflected in the moral priorities of existing international human rights regimes. Within predominant contemporary diagnoses of, and prescriptions for, world politics the problem is not that the co-existence of a plurality of orientations goes unrecognised, so much as that the meaning of this plurality is always already homogenised by reference to the authoritative space/time of western modernity. It is the subjective certainty of this orientation that not only grounds the theorist's judgment but also enables it to make a difference in practice, through timely prescription and through example. Instead of being the one who already knows the time, the heterotemporally oriented theorist is fundamentally uncertain of his own punctuality (see Thaler 2014 for an attempt to complicate temporal assumptions in just war theory). The extent to which his interventions are or are not timely will depend on the moral/temporal certainties and uncertainties (orientations) of his interlocutors. To engage with alternative temporal framings for judgment is well within the limits of logical possibility, but to take seriously a challenge to one's investment in a narrative of truth and progress that cannot live with provincialisation, is profoundly disturbing for those of us educated in rationalist traditions of moral theorising. It is, however, the only way to shift the ground of ethical debate about world politics away from an agenda that is incapable of seriously questioning its own timeliness.

References

Chakrabarty, D. (2000) *Provincialising Europe.* Princeton: Princeton University Press.

Fabre, C. (2012) *Cosmopolitan War.* Oxford: Oxford University Press.

Hutchings, K. (2008) *Time and World Politics: Thinking the Present.* Manchester: Manchester University Press.

Hutchings, K. (2010) *Global Ethics: An Introduction.* Cambridge: Polity.

Hutchings, K. (2011) 'What is Orientation in Thinking? On the Question of Time and Timeliness in Cosmopolitical Thought.' *Constellations: An International Journal of Critical and Democratic Theory* 18 (2): 190-204.

McMahan, J. (2009) *Killing in War.* Oxford: Oxford University Press.

Miller, D. (2008) 'Political Philosophy for Earthlings.' In: Leopold, D. and Stears, M. (eds.) *Political Theory: Methods and Approaches.* Oxford: Oxford University Press.

Mills, C. (2005) 'Ideal Theory" as Ideology.' *Hypatia: journal of feminist philosophy* 20(3): 165-84.

Simmons, J. (2010) 'Ideal and Nonideal Theory.' *Philosophy and Public Affairs* 38(1): 5-36.

Singer, P. (1972) 'Famine, Affluence and Morality.' *Philosophy and Public Affairs* 1(3): 229-43.

Thaler, M. (2014) 'On Time in Just War Theory: from *Chronos* to *Kairos.*' *Polity* 46(4): 520-546.

3

The Epistemological Consequences of Taking Time Seriously and the Value of Generational Analysis in IR

TIM LUECKE
OHIO STATE UNIVERSITY, USA

As the very existence of this volume demonstrates, scholars of International Relations (IR) are increasingly paying attention to the importance of time and how time affects processes and outcomes in international politics. In this chapter, I will not make the case for why we should incorporate time into our analysis, but assume that we can take it for granted that it is an endeavour worthwhile pursuing. Once we assume that there is a case for taking time seriously, two questions pose themselves. First, how do we actually take time and the temporality of international politics more explicitly into account, and incorporate time into our existing theoretical frameworks? Second, what are the epistemological implications of incorporating time into the ontology of our theoretical frameworks and how do we address them in our research.[17] Pointing out that taking temporality seriously requires us to rethink our epistemological commitments is one thing, showing how to actually incorporate those epistemological implications is another.

In the pages below, I will argue that one particularly promising way to incorporate the temporal dimension of international politics into our theoretical apparatuses is to adopt a generational method of analysis. The 'generation' in essence constitutes a temporal unit of analysis, which locates individuals and

---

[17]   Loosely defined, while 'ontology' is concerned with what the world is 'made of', 'epistemology' is concerned with how we know that world.

collectives in the process of time. A generational method therefore allows us to study how groups that are located at different positions in time interact and consequently affect political outcomes.

My answer to the second question is that adopting a generational method and taking the temporality of political ideas and practice more seriously implies that we, students and scholars of international politics, cannot claim to occupy a timeless position from which to evaluate international relations and foreign policy. Taking time seriously implies that the knowledge we produce as a discipline is always time-dependent, since we always interpret the world from the temporal position of our own generation. This implies the possibility that the validity and 'truth' of our research varies historically. One could consequently question whether our research and the knowledge we produce can therefore be judged and evaluated in an objective fashion. This threat of relativism, however, can be countered by making the observer's position, i.e. our own position, in time explicit and by incorporating it directly into our own research. I show that generational analysis helps us to take a first step in addressing the epistemological implications of taking time seriously by allowing us to study our own perspective in the temporal process via an analysis of our own generation and how it relates to the people whose actions and ideas we study.

In order to illustrate how to apply generational analysis and how to address the epistemological consequences outlined above, I will discuss my own research on the role of the WWII Generation in post-war Germany. I have only recently begun this project, which builds on my dissertation, and the pages to follow will consequently help me to think through some of the issues that I will have to address in this project. You should therefore consider this an initial 'think piece' which contains a lot of ideas that will need to be developed much more carefully in the future. At the same time, however, I hope that drawing on my own work will not only show the complexities of dealing with time in our research, but also provide a somewhat concrete example of how one can possibly address these complexities.

Temporality and Generations

Time is, alongside space, one of the most fundamental conditions of our existence. We are temporal beings in so far as we are born, age, and die, and we experience the social world around us as a constant stream of events that signify the passage of time. In order to understand how temporality affects outcomes in international politics, we therefore ideally possess theoretical frameworks that are dynamic, by which I mean that they are capable of incorporating changes and processes that occur over time. Most theories in

IR, however, are static and a-temporal, in so far as they focus on the structure of incentives, the international system at any given point in time, etc.[18] In my own work, I have tried to demonstrate that generational analysis provides one powerful complement to our theoretical toolbox that enables us to take the temporal nature of international politics and foreign policy into account more explicitly.

In this section I will explain how the concept of generations and generational analysis provide one particularly fruitful way to take the temporality of our 'objects of study,' and therefore the temporality of foreign policy and international politics, more seriously. By 'temporality' I mean the experience of time (McIntosh 2015: fn 2), and we experience time through the passage of events, our actions in the present, and our expectations for the future. The analytical value of the concept of 'generations' is that it locates individuals and groups at the intersection of two of the most fundamental temporal dimensions, collective and individual time. Collective time refers to events which are experienced by society as a whole. Elections, national sports tournaments, or, most extremely, war, are just some examples of such collective events. *How* we experience these events and therefore our time, however, also depends on the particular life stage we occupy, or on our individual time. Events experienced during childhood will have a different effect on someone than events experienced as an adult. Most generation scholars agree that the age of youth plays an especially important stage in life given that this is the time when most individuals develop and solidify their basic political orientations and worldviews.[19]

The generation then signifies the intersection between individual life stage and collective events. As such, it constitutes a unit of analysis that locates people in the process of time, similarly to how the concept of 'class' locates people in the existing social structure. However, in contrast to analytical concepts, such as social class, the state, or the international system, all of which capture a particular aspect of social reality at any given *point* in time, the concept of generations constitutes a temporal unit of analysis which captures the different experiences of groups of people located at different intersections of individual and collective time. This implies that different generations live, quite literally, in different times because 1) they experienced a particular set of events during a different life stage than their predecessors or successors, or because 2) they experienced different sets of events during

---

[18]   See, for example, McIntosh (2015).

[19]   On the relevance of the age period of youth and for good general introductions to the concept of generations and its explanatory contribution see, for example, Eisenstadt 2003 (1956); Edmunds and Turner (2002, 2005; Fogt (1982); and Mannheim 1997 (1952). For a discussion of the concept of generations and its application to foreign policy analysis and the study of IR, see Jervis (1976) and Steele and Acuff (2012).

the same life stage. Importantly, this means that different generations share a different worldview and attach different meanings to the ideas that they inherit from preceding generations. 'Democracy' does mean something different to me than it did to my grandparents. Generational analysis allows us to study the interaction of differently positioned generations and how those interactions affect political outcomes.

In contrast to a linear conception of time, where time moves forward as a singular stream of events, a generational perspective implies a conception of time which is comprised of multiple temporal experiences, or generations, which co-exist simultaneously while still living in different 'times'.

The potential range of applications of generational analysis to the study of foreign policy and international relations is enormous, given that most of the processes we study take place, by definition, over the course of time. Instead of going through a list of applications of a generational approach,[20] I will focus on the concept of political generations and use my own work as an example to illustrate how to actually apply generational analysis. This will also set up my argument that adopting a generational approach, and more generally taking temporality seriously, has important implications for how we study international politics and for the status of knowledge that we produce in the discipline.

Political Generations

In my work I have drawn on the concept of generations in order to develop a theory of political generations and to show that political generations have the potential to explain periods of change and stability in foreign policy and international politics. I define political generations as cohorts in their youth who, in response to a series of events that seem to challenge the existing social and political order, develop a sense of generational consciousness and belonging.[21] As a result, they develop distinct, historically specific, sets of political worldviews and act in the world on the basis of shared background beliefs that make political practice and discourse possible in the first place.

---

[20]   See Steele and Acuff (2012) for an excellent showcase of the potential breadth of applications of generational analysis. Also, for a much more modest attempt, see Luecke (2013b).

[21]   Note that only events which 'seem to challenge the established political' qualify as events capable of resulting in new political generations. For example, while the war in Korea was not perceived as challenging the political order of the U.S. at the time, the war in Vietnam certainly was perceived as such a challenge by both the political right and left. In the end, whether or not a particular series of events results in the emergence of a new political generation is an empirical question.

The fact that generational experiences are shared by entire cohorts, however, does not imply that the members of a generation agree on how to respond to the challenges facing their society. Quite the contrary, members of one and the same generation might disagree heavily how to address those challenges. What makes a generation a political generation is that its members have to deal with the same set of historical circumstances at a similar stage in their life cycles. The members of political generations identify with one another through the challenges that formative events pose, but they often disagree heavily on how to respond to those challenges.

Political generations can be more 'radical' or 'traditional', depending on whether their members perceive the old generation in power to respond to the challenge facing society successfully or not. If elites, and therefore the old generation, are perceived as failing to address the challenge, members of the newly emerging generation will attempt to change the existing political culture. The Sixties Generation is maybe the most famous example of such a radical political generation. However, as the fate of the Sixties Generation has shown, attempts by the young generation to change the established order are often doomed to failure. The reason for this is that power is largely concentrated in the hands of the older generation, which will often attempt to resist the challenges of the young generation. However, even if political generations fail to bring about change at the time of their 'birth', their members eventually age and start to replace the older generation from positions of power. One of the most interesting and potentially valuable insights of a theory of political generations is therefore that the initial cause of change, a series of formative events, and its effect do not follow immediately upon each other, but only occur with a time lag of roughly 15-25 years.[22] A generational approach is therefore able to provide new answers to old questions, especially when it comes to investigating the long term effects domestically and internationally of events such as World War I and II, Vietnam, the attacks on September 11, 2001, the Arab Spring, or the economic crisis that started in 2008.

Whereas generational change that brings to power a radical generation is likely to result in changes in political culture and political practice, the coming

---

[22]   If we define the age of 'youth' as roughly the age of 18-25 and consider that most people in positions of power are at least 35 years of age or older, we can project that generational change will begin about 15-25 years after the generation experiences its formative events. In general, the boundaries of political generations are always slightly fuzzy since even individuals who do not fall exactly into the age bracket defined as formative might identify with the formative experiences of a generation. However, most concepts we employ in IR, such as the state or culture, have fuzzy boundaries that are difficult to define with precision, which does not stop us from drawing on them in our research.

to power of traditional generations is marked by stability. Traditional political generations emerge when cohorts in their youth perceive the response of the old generation to challenges facing the political order as successful. Instead of trying to radically change the existing political order, traditional generations therefore adopt the political culture of the old generation, even though they might adjust it to new circumstances.

Political generations therefore constitute one potentially powerful mechanism to adjust political culture, and by extension ideas and beliefs about foreign policy and international politics, to an ever changing world. Based on this argument, I suggest that radical and traditional political generations alternate across time, thereby giving rise to cycles of change and stability in foreign policy and international politics.[23] In my dissertation, I subject this theory of political generations and generational change to an initial plausibility probe through an examination of U.S. foreign policy and the emergence of the West in the 20th century. While not conclusive, the results show that the timing of generational changes matches the timing of periods of change and stability in U.S. foreign policy and international politics.

Epistemological Implications

The argument that I have tried to make so far is that generational analysis can provide us with access to the temporality of the people whose ideas and practices we study in the context of foreign policy and international politics. However, adopting a generational perspective and thereby incorporating temporality into the ontology of our theoretical frameworks also has important epistemological implications, or put differently, implications for how we can know, understand, and consequently study the world. Namely, if different generations experience the world differently and therefore interpret the world differently, then we, as researchers, cannot occupy a 'timeless' position from which to study those generations, since we ourselves occupy a particular generational location in the process of time. This implies that the knowledge that the discipline of IR, and all social science for that matter, produces cannot be a-temporal and 'objective', but is always temporal itself, time-bound, and contextual.[24] The contribution of a generational approach to the study of foreign policy and IR, however, rests in not only alerting us to the epistemological implications of taking time seriously, but primarily in providing a first step in addressing these implications in our research.

Let me briefly illustrate through my own current research how this problem manifests itself in research practice and how generational analysis provides a

---

[23] For a more detailed discussion see Luecke (2013a).

[24] I am certainly not the first one to make this claim. See especially Kratochwil (2006).

solution to said problem. About a year ago I began working on a new research project, which investigates the role of the WWII Generation, those who experienced the Second World War during their formative years of youth, in post-war Germany. I am German myself and the question that made me study IR and enter academia was 'how is it possible that a country, such as Germany, starts two world wars, brings incredible destruction to the world, and then changes into one of the most stable, economically successful, and peaceful nations on earth?' Obviously there already exists a wide range of answers to this question, provided by realist, liberal, or constructivist theories. However, most of these answer focus on external factors, such as the occupying forces, the balance of power, etc. Yet, few accounts in IR that I am aware of have examined the role of Germans and especially of those Germans who had fought during the war.[25]

However, there is also a more personal motivation to pursue this project. My grandfather, Paul Lücke, had been minister of housing and for a brief period minister of the interior for the Conservative Democratic Union (CDU) during the post-war years in Germany. He died before I was born and since we had never spoken much about him in my family, I had planned to write a biography for some years already. Paul Lücke was born in 1914 and had fought as a soldier of the Wehrmacht during the entire course of the war from 1939 and 1945. Despite his participation in the war, he had never joined the Nazi party and actively opposed the Nazis during his youth on the basis of his Christian-Conservative values. Those values also informed his latter public housing policies and his efforts to house the millions of refugees that were returning to Germany after the war. His biography therefore provides a perfect starting point and analytical foil to examine the contribution of members of the WWII Generation who were not convinced Nazis and who contributed significantly to the successful reconstruction and re-integration of Western Germany. For that reason, I decided to channel both these motivations into a single project.

However, in the context of discussing this project with colleagues, friends, and a very observant archivist at the Konrad Adenauer Foundation, I was asked repeatedly whether the relationship to my grandfather would not affect my interpretation of history and make it impossible to remain objective and unbiased. I was obviously aware of the issue already and knew that I had to address the question. However, I also quickly noticed that I needed to address not only the question of my personal relationship to my grandfather, but my relationship to his generation as a whole. The WWII Generation obviously has a particular public image in Germany, given that many of its members were implicated in the Nazi regime, the Holocaust, and other mass

---

[25]   Jackson (2006) is a notable exception.

atrocities. Moreover, this public image of the WWII generation in Germany has clearly been shaped by the public discourse of another political generation, the Sixties Generation, or the '68er Generation' (Generation of 1968), as it is best known in Germany. The Generation of 1968 emerged, just like in the United States, in large part as a response to the war in Vietnam, racial unrest, the perception that liberal Western countries were supporting repressive regimes in Third World countries, but also in response to the fact that many former Nazis or Nazi supporters were still occupying positions of public power in Western Germany. Unquestionably, the Generation of 1968 and the fact that it contributed significantly to an actual engagement with the past and the crimes that were committed by those who started and fought in the Second World War has been a positive influence and a significant reason for the political and cultural change that has made Germany the country that it is today.

At the same time, the public discourses and images of the WWII Generation that the Sixties Generation created in Germany, and which still dominate the political discourse today, have also resulted in the fact that the contribution of members of the WWII Generation who were not Nazis has often been ignored or 'forgotten'. Now, I myself was raised and socialised by members of the Generation of 1968. Not only my parents, but my teachers, high school teachers, and many adult role models during my youth belonged to the Sixties Generation. As a result, my political worldview and also my perspective on the WWII Generation are already shaped and biased by the interpretations of the Generation of 1968. The complicated relationship between the WWII Generation, the Generation of 1968, and my own generation therefore already makes it difficult to see how an 'objective history' of the WWII Generation is possible, and it shows that taking temporality seriously does come at the cost of increasing complexity and intricate epistemological problems.

How should one deal with the problem that there seems to be no timeless, and therefore no objective, position from which to evaluate the contributions of the WWII Generation in Germany? I think that the only defensible course of action is to confront this problem head on and to explicitly address and incorporate the temporal nature of my position and consequently of the results that I generate into the analysis. Put differently, we should not merely accept, but embrace the fact that we evaluate the past and try to predict the future from a particular perspective in time which will inevitably affect the results of our research. Generational analysis provides a clear answer to the question of how to uncover this position in time; namely, by studying our own generation and the particular worldview and formative challenges that characterise it. In the case of my research project, which tries to examine the role of the WWII Generation in Germany, I will therefore not only examine the political orientations of the WWII Generation, but also those of the Generation of 1968, and of my own generation. More importantly, I will need to disentangle the complicated ways with which these three generations relate to one another.

The Threat of Relativism

Taking time seriously, for example by studying generations from one's own generational location, means to accept that there is no 'objective' or 'timeless' position from which to study social life. Even though this is a discussion that I can clearly not engage in much detail here, the lack of a timeless position runs contrary to the tenets of positivism, which still dominates the field of IR and it raises the question of the status of knowledge thus generated.

Indeed, it is no coincidence that the author of the most seminal essay on generations, Karl Mannheim, was first and foremost famous for his attempts to develop a sociology of knowledge, which he understood to consist of a 'theory of the social or existential conditioning of knowledge by location in a socio-historical structure' (Pilcher 1994:482). Much of Mannheim's work rested on a critique of the Enlightenment idea that reason or truth are temporally static. Instead, he argued that reason is historically dynamic (Hekman 1986: 54), or put differently, the criteria for evaluating good arguments and by extension good research vary across time. This is much in line with the thought of the philosopher of science Thomas Kuhn (1996), who argued that science progresses not in linear fashion but in paradigm shifts which (coincidence?) are caused by generational changes.

The critique that Mannheim encountered, and which bedevils any attempt at a sociology of knowledge more generally, is that his arguments ended up in relativism, the idea that there is no way at all to differentiate between 'good' and 'bad' reasons, research, or 'truths'. If reason varies historically and if there are therefore no objective criteria with which to evaluate knowledge, isn't all knowledge and therefore also all IR research entirely up to our, or in this case my own, subjective standards?

My answer to this question is a cautious and preliminary 'no'. Yes, knowledge from a particular generational location or temporal perspective is not 'objective knowledge', understood as knowledge from some ultimate timeless Archimedean viewpoint. Yet, it is neither merely subjective. Instead knowledge, and therefore the research we produce, is intersubjective in so far as it rests on standards of evaluating reasons that are shared in the community that is 'IR'. More generally, the knowledge and practices we produce and engage in can be evaluated by members of our generation, who share the same temporal location and therefore similar standards of evaluation. In addition, once we make our own generation location explicit, our research can be assessed by older or younger generations, since they will be able to locate themselves vis-à-vis our generation in the historical process and thereby be able to understand the particular context in which our research was generated.

Conclusion

As I already stated up front, this is only a very first cut at the question of how to address the epistemological implications of taking time seriously. This question requires a much more in-depth discussion and a careful engagement with the relevant literature in social theory, sociology of knowledge, and obviously IR. My primary goal in the pages above was to take a first initial stab at thinking through the consequences of taking temporality seriously and to use my own work as an example for how one can potentially deal with those issues. I hope that I have been able to show that generational analysis is one promising way to incorporate the temporality of international politics into our theoretical frameworks. However, adopting the notion that generational membership shapes the way we perceive, interpret, and act in the world implies that we ourselves, the knowledge we produce, the practices we engage in, are shaped by our generation, which is the location we occupy in the temporal process. I think that this requires us to explicitly incorporate our own temporal location into our research and accept the fact that our results are judged by community standards that are upheld and changed by generations of IR scholars and students. This argument therefore comes to similar conclusions to those of Friedrich Kratochwil, who argued that history understood as memory 'is always viewed from a particular *vantage point of the present*' (Kratochwil 2006: 21). Kratochwil arrives at this conclusion through a discussion of the role of history and the ability of the discipline to generate practical, rather than merely theoretical, knowledge. The fact that we seem to arrive at similar conclusions, however, suggests that if we want to take time seriously in IR research we will actually need to think through the epistemological implications and complications that are entailed in making such an analytical shift. In my opinion it will certainly be worth the costs. Yet, developing these ideas further will require a collective effort. I therefore hope that the next generation of IR students and scholars will take up the challenge.

* The ideas developed here benefited greatly from discussions with Lise Herman and Alexander Wendt. My thanks and gratitude go out to both of them. All mistakes and errors are obviously my own.

References

Edmunds, J. and Turner, B. (2002) *Generations, Culture and Society.* Philadelphia: Open University Press.

Edmunds, J. and Turner, B. (2005) 'Global Generations: Social change in the Twentieth Century.' *The British Journal of Sociology* 56(4): 559-577.

Eisenstadt, S.N. (2003[1956]) *From Generation to Generation.* New Brunswick: Transaction Publishers.

Fogt, H. (1982) *Politische Generationen: empirische Bedeutung und theoretisches Modell.* Opladen: Westdeutsche Verlag.

Hekman, S. (1986) *Hermeneutics and the Sociology of Knowledge.* South Bend: University of Notre Dame Press.

Jackson, P. (2006) *Civilizing the Enemy: German Reconstruction and the Invention of the West.* Ann Arbor: University of Michigan Press.

Jervis, R. (1976) *Perception and Misperception in International Politics.* Princeton: Princeton University Press.

Kratochwil, F. (2006) 'History, Action and Identity: Revisiting the 'Second' Great Debate and Assessing its Importance for Social Theory.' *European Journal of International Relations* 12(1): 5-29.

Kuhn, T. (1996) *The Structure of Scientific Revolutions.* Chicago: University of Chicago Press.

Luecke, T. (2013a) 'Generations in World Politics: Cycles in U.S. Foreign Policy, the Construction of the "West," and International Systems Change, 1900-2008.' PhD thesis, Ohio State University.

Luecke, T. (2013b) 'The Nexus of Time: Generations, Location of Time, and Politics.' *Qualitative & Multi-Method Research* 11(2): 9-14.

Mannheim, K. (1997 [1952]) 'The Problem of Generations.' In: Hardy, M.A. (ed.) *Studying Aging and Social Change*. London. U.K.: Sage Publications. 22-65.

McIntosh, C. (2015) 'Theory Across Time: The Privileging of Time-Less Theory in International Relations.' *International Theory*, *FirstView*. Available online at: http://journals.cambridge.org/action/ displayAbstract;jsessionid=CD0EF8F211BCC79531CAB3BAADB66B63. journals?aid=9970859&fileId=S1752971915000147

Pilcher, J. (1994) 'Mannheim's Sociology of Generations: An Undervalued Legacy.' *The British Journal of Sociology* 45(3): 481-495.

Steele, B. and Acuff, J. (eds) (2012) *Theory and Application of the 'Generation' in International Relations and Politics.* New York: Palgrave Macmillan

4

Governing the Time of the World

TIM STEVENS

ROYAL HOLLOWAY, UNIVERSITY OF LONDON, UK

Recent scholarship in International Relations (IR) is concerned with how political actors conceive of *time* and experience *temporality* and, specifically, how these ontological and epistemological considerations affect political theory and practice (Hutchings 2008; Stevens 2016). Drawing upon diverse empirical and theoretical resources, it emphasises both the political nature of 'time' and the temporalities of politics. This *chronopolitical* sensitivity augments our understanding of international relations as practices whose temporal dimensions are as fundamental to their operations as those revealed by more established critiques of spatiality, materiality and discourse (see also Klinke 2013). This transforms our understanding of time as a mere backdrop to 'history' and other core concerns of IR (Kütting 2001) and provides opportunities to reflect upon the constitutive role of time in IR theory itself (Berenskoetter 2011; Hom and Steele 2010; Hutchings 2007; McIntosh 2015).

One strand of IR scholarship problematises the historical emergence of a hegemonic global time that subsumed within it local and indigenous times to become the time by which global trade and communications are transacted (Hom 2010, 2012). This linear and mechanical time evolved in lock-step with industrialisation and the globalisation of capital and through the internet, in particular, a temporal infrastructure has emerged that supports the greater infrastructures of global exchange across a range of massively distributed yet tightly interdependent sociotechnical systems. This form of time-reckoning is contrasted with the complex *heterotemporality* of actual human lives under conditions of late modernity (Hutchings 2008: 1172-6) but relatively little attention has been directed towards the nature and character of global time itself. Terms like 'Western standard time' have served more as placeholders for critique than objects of empirical analysis in their own right (Hom 2010: 1169). This chapter attends to this analytical deficit by asserting the internal heterogeneity of 'global time' but at the same time describing how the *global governance* of time seeks to render this temporal assemblage unified, meaningful and useful. It asks the simple question: who governs the time of the world? Who and what is responsible for producing the hypermodern, technological *chronos*, the 'always-synchronised time of life, society and world history'? (Stevens 2016: 188) The focus of this chapter is the production of Coordinated Universal Time (UTC), the foundation of global broadcast and civil time since 1961.

The first section of this chapter explains what is being governed and why. A brief historical review establishes the context and development of UTC and shows how the *times of the universe* are translated into the *time of the world* for the purposes of global synchronisation and communication. The second section identifies the *key actors and mechanisms* involved in UTC and therefore in the global governance of time. This includes a range of intergovernmental organisations, non-state actors and expert networks. The third section introduces some issues and problems arising from this form of global governance, particularly those surrounding 'leap seconds'. The chapter concludes by suggesting possible avenues of enquiry for IR and global governance scholars concerned with the political construction and contestation of time and temporality in the contemporary world.

A Brief History of Time

Coordinated Universal Time (UTC) is the basis for worldwide broadcast and civil time. It informs multiple time scales from Global Positioning System (GPS) time and internet time protocols to radio time signals and national speaking clock services. It is the most widely used global time scale and underpins every imaginable human activity relying upon precise time reckoning at the local, national and global levels. The production of UTC demonstrates that global time is not a matter of simply 'reading off' time from the universe but is a process of assembling and negotiating different ways of calculating and thinking about time. This section outlines the essential scientific dynamics of this process; the subsequent section will explore in greater detail the actors responsible for the production of UTC and the global governance of time.

The foundation of UTC is the second, the base unit of time in the International System of Units (SI) and the basis of all modern time-reckoning and measurement. Since at least the 15th century, seconds, minutes and hours have been calculated as divisions of the solar day, itself derived from astronomical observations of the Sun 'rising' or 'setting' relative to fixed terrestrial markers. The precise length of the solar day changes due to geological, astronomical and relativistic factors; units derived from its length have therefore always been somewhat approximate. Nevertheless, the calculation of the *mean solar second* as 1/86,400th of a mean solar day not only persisted into the twentieth century, but in 1935 was adopted by formal international agreement as the fundamental scientific unit of time (Kennelly 1935).

By 1928, the mean solar second was informing Universal Time (UT), a global time scale therefore also derived from the axial rotation of the Earth relative to the Sun. The recognition that the length of the solar day fluctuated led to the development of new time scales that took account of factors like atmospheric drag, the gravitation of the Moon and a slow but discernible deceleration in axial rotation speed.[26] One such was Ephemeris Time (ET), adopted in 1952 by the International Astronomical Union, and calculated not on the length of the solar day but on the duration of the Earth's annual transit around the Sun (i.e. the year). In 1960, the 11th General Conference on Weights and Measures (CGPM), which adjudicates on international metrological issues, further refined this 'ephemeris' second as a fraction of the length of the baseline year 1900. This marked a formal shift from using observed solar time to derive the SI second – and, therefore, global time scales founded on the second – to its definition through Newtonian celestial mechanics and, subsequently, by other techniques that accounted for the effects of relativistic motion. All were calculated with reference to physical and observable objects amenable to visual identification and measurement.

In 1967, the 13th CGPM adopted the *atomic second* as the SI base unit of time. Advances in atomic clock technology presented more precise means of establishing consistently the duration to be known as a 'second'. Global time was tethered for the first time not to the observable passage of astronomical objects but to invisible radioactive events, specifically the frequency of oscillations of the caesium-133 atom (Audoin and Guinot 2000).[27] The new International Atomic Time (TAI) surpassed astronomical time in accuracy but for reasons of continuity with the existing system the specific qualities of the atomic second were chosen to match that of the ephemeris second. However, these two values have since drifted apart, not least due to the persistent slowing of the axial rotation of the Earth. The challenge for the new

---

[26]   The following account is drawn principally from Nelson *et al.* 2001.

[27]   The second is defined in the SI system as 'the duration of 9 192 631 770 periods of the radiation corresponding to the transition between the two hyperfine levels of the ground state of the caesium 133 atom' (BIPM 2006: 133).

Coordinated Universal Time (UTC), introduced in 1961, was to harmonise astronomical time (UT) with atomic time (TAI) to produce a time scale (UTC) that could be used by all the disparate communities dependent upon accurate time measurement. Then as now, laboratory-based sciences preferred atomic time—on account of its regularity and reliability—whereas spatially dependent activities like astronomy and navigation required time derived from solar and other forms of astronomical time-reckoning. The very existence of UTC can be read as an attempt to reconcile divergent disciplinary requirements under a unifying temporal rubric.

Various experiments were undertaken throughout the 1960s to implement irregular 'jumps' or 'time steps' of fractions of seconds in TAI to align it with UT and therefore to be harmonised within UTC. Stepped Atomic Time (SAT), for instance, 'ticked' at an identical rate to TAI but interpolated jumps of 200 milliseconds to keep it synchronised with UT. Measures like SAT failed to resolve the situation and global time continued to be reliant on two 'seconds' of slightly different length, one of whose length also varied. Proposals emerged that all UTC and TAI seconds should correspond to the SI definition and that any changes to UT should be via the addition or subtraction of integer (whole) seconds to keep astronomical and atomic time coordinated. On 1 January 1972, this system of 'leap seconds' was inaugurated formally and continues to keep the offset between UT and TAI to an integer value that makes the calculation of UTC more straightforward and respectful of both atomic and solar time scales.[28] Leap seconds have been described as a 'crude hack' (Kamp 2011) in the structure of time and are in effect a metrological response to the heterotemporality of technological time-reckoning. One fundamental distinction is between times derived from atomic sources and those from astronomical observation, which, despite their superficial similarity, are quite different qualitatively and quantitatively. The foregoing abridged description does not exhaust this heterotemporality, as many other time scales exist, but three key observations emerge from the brief discussion of UTC alone. First, there is no single global time scale for the unequivocal reckoning of time. This is because of the relativistic nature of time itself: what we think of as objective time cannot simply be 'read off' a physical universe in which all time is local if it exists at all (Barbour 1999). The atomic second was mapped to the solar second, not the other way around, and, as the two diverge in quantity, ways must be found to account for and accommodate those changes. This leads to the second point, that global time – in this case, UTC – is a matrix of competing times based on different calculations, techniques, material objects and practical imperatives that must be negotiated and, ultimately, governed. The third observation is that

---

[28]   This is summarised by the basic formula, UTC = TAI – leap seconds. The leap seconds ensure that UTC (essentially an atomic time) never differs from solar (or rotational) time (UT1) by more than 0.9 seconds.

temporal or, perhaps better, *global chronometric governance* is enacted in and through assemblages of actors and institutions. Who or what these are, and how they do it, is the topic of the following section.

Actors and Mechanisms in Global Chronometric Governance

Space precludes an extended discussion of global governance but its central intended meaning in IR is the description and explanation of the multiple activities that result in coordinated political action at the global scale, in the absence of a world government (Rosenau and Czempiel 1992). The emergence of global governance as an analytical lens coincided with globalisation and the end of the Cold War and indicated a concomitant commitment to exploring new ways of ordering in a post-bilateral world (Zumbansen 2012: 84). Historically, however, global governance as political *practice* has a longer heritage than its recent analytical introduction suggests. The global governance of time, for instance, dates to the 1870s and 1880s and to the first international attempts to harmonise and standardise scientific and technological ways and means on a global scale (Mazower 2012). The global governance of time *pre*-dates many other forms of global governance and, indeed, has made many of them possible. These early attempts to govern global time were contested, for reasons of national pride as much as disagreement over technical standards and scientific accuracy. Having failed to convince the British and Americans to adopt the metric system of measurement it developed, France abstained from voting for the adoption of Greenwich Mean Time at the International Meridian Conference (1884). Paris Mean Time would instead persist into the twentieth century, an awkward nine minutes and 21 seconds out of step with the rest of the world (Palmer 2002). The governance of time has always been a key aspect of modernity and continues to be highly political, if not always rendered as such.

Specifically, the study of global governance has four principal concerns: the global nature of contemporary problems and their potential solutions; the agency of non-state and transnational actors; the conception of 'order' as a dynamic phenomenon having diverse foundations other than political-legal authority; and, a normative concern with how best to effect positive sociopolitical change (Hofferberth 2015: 601). These four aspects are discernible in the discussion of global chronometric governance. Its essential purpose is to translate the unruly localism of all possible times into one unifying time with which the world can transact its business. It does this through state and non-state actors acting transnationally across borders and continents. The order so derived is contingent principally on domains of scientific knowledge rather than political-legal authority and is informed by normative concerns, the contestation of which will be discussed in a later section.

This section will focus on what Avant *et al.* (2010) call 'global governors'. These are 'authorities who exercise power across borders for purposes of affecting policy', and who 'create issues, set agendas, establish and implement rules of programmes, and evaluate and/or adjudicate outcomes' (Avant *et al.* 2010: 2). This attention to agents and agency within global governance structures emphasises the constructed and dynamic nature of global governance, in which '[n]othing is ever governed once and for all time' (Avant *et al.* 2010: 17) but is subject to iterative processes of mediation and negotiation. Whilst this is true of chronometric governance too, the persistence of many of the key actors is striking. For instance, The *Convention du Mètre* (Metre Convention) of 1875 brought into being three intergovernmental organisations to coordinate international metrology and internationalise what was originally the European metric system. Since 1960, they have also been responsible for maintaining the International System of Units (SI). The senior decision-making body is the General Conference on Weights and Measures (CGPM), which meets every four to six years to discuss and adjudicate on significant metrological issues. One instance was mentioned previously, the adoption of the atomic second by the 13th CGPM in 1967. More recently, the 25th CGPM discussed but did not decide on the redefinition of the kilogram, the last remaining SI base unit derived from a physical artefact rather than a physical constant (Karol 2014). The CGPM is advised by an expert panel that meets annually, the International Committee on Weights and Measures (CIPM), which also informs the work of the third organisation, the International Bureau of Weights and Measures (BIPM), of most concern here.

One of BIPM's many roles is to ensure the proper administration of Coordinated Universal Time (UTC), although the technical specifications for doing so are defined by the International Telecommunications Union (ITU), a specialist agency of the United Nations. This requires that the BIPM produces a time scale accurate to within one-tenth of a second (ITU 2002). The ITU fulfils this slightly incongruous role on account of its historical status as the regulator of the shortwave radio networks through which global time scales were disseminated prior to satellites and the internet (Beard 2011). Like the BIPM and its peer organisations, the ITU has its origins in 19th-century international attempts to regulate scientific and technological issues and its earlier incarnation, the International Telegraphy Union, should probably be considered one of the first public international unions (Mazower 2012: 102). It was founded in 1865 and is still the world's premier forum for the global governance of information and communications technologies, even if it no longer enjoys universal support.

BIPM produces UTC from a combination of International Atomic Time (TAI) and Universal Time (UT). As described previously, UT is defined by the rotation of the Earth, based originally on astronomical observations but now calculated on measurements provided by satellites. Since 2003, the principal form of UT has been UT1, produced by the International Earth Rotation and Reference Systems Service (IERS), the various components of which are distributed across the US, Europe and Australia.[29] The IERS Sub-bureau for Rapid Service and Predictions of Earth Orientation Parameters located at the US Naval Observatory (USNO) in Washington, DC, publishes a weekly forecast of the next year's daily values of UT1 in its weekly Bulletin A. These values are derived from data obtained by a variety of large-scale observational techniques, including Very Long Baseline Interferometry (VLBI), GPS satellites, Satellite Laser Ranging (SLR) and Lunar Laser Ranging (LLR), which requires collaboration with multiple scientific institutions and organisations.[30] IERS also publishes Bulletin C, which provides six months' notice of the introduction of leap seconds. Since leap seconds are principally to keep the offset between TAI and UT1 to an integer value, Bulletin C is the primary document of concern to the BIPM's calculation of UTC.

UT1 gives UTC the 'time' we recognise intuitively, calculated from the rotation of the Earth, but atomic time (TAI) gives it stability. The initial step in calculating TAI is the production of 'free atomic time' (Échelle atomique libre, or EAL), which is obtained from over 420 atomic clocks based at 72 time laboratories on every continent except Antarctica.[31] These communicate with each other via a complex array of navigation and communications satellites, a 'time transfer' system managed by BIPM but locally calibrated by various actors, including manufacturers of atomic clock technologies. TAI is calculated by comparing EAL to primary frequency standards calculated by atomic clocks like the NIST-F1 at the US National Institute of Standards and Technology (NIST) laboratory in Boulder, Colorado.[32] This newest generation of atomic clocks is accurate to fractions of a second over geological time scales and continues to improve. If EAL drifts from these standards, a correction is applied at the level of nanoseconds and TAI is produced. The final step in calculating UTC involves the addition of leap seconds from IERS Bulletin C to UTC when necessary. At present, each terrestrial day is about 2.5 milliseconds longer than the last, while atomic time remains constant. This incremental lengthening equates to approximately one second per year, requiring the addition of a leap second every one or two years, always in June or December.

---

[29]   http://www.iers.org/IERS/EN/Home/home_node.html.

[30]   http://www.usno.navy.mil/USNO/earth-orientation/eo-info/general/input-data/input-data-series-used-in-iers-bulletin-a.

[31]   http://www.bipm.org/en/bipm/tai/tai.html.

[32]   http://www.nist.gov/pml/div688/grp50/primary-frequency-standards.cfm.

The resulting UTC values are disseminated in BIPM's monthly Circular T, which allows participating laboratories to adjust their local time scales. A curiosity of the system is that it is not possible to know the precise value of UTC at any given time or place: there is no absolute value or physical artefact to which reference might be made. Local values known as UTC(k) can be tracked at particular laboratory sites but UTC is only approximate on account of its compiled nature. Since 2013, the BIPM has released a weekly 'Rapid UTC' solution (UTCr), to enable more frequent steering of local atomic clock times, with an error relative to UTC of less than two nanoseconds. BIPM also publishes TT(BIPM), a realisation of Terrestrial Time defined by the International Astronomical Union (IAU) and used for location-dependent astronomical purposes.

A picture emerges that demonstrates the constructed nature of global time and the complex actor-networks that enable its production. These include intergovernmental organisations, expert groups, national and independent laboratories, international scientific collaborations and equipment manufacturers. Crucially, they also rely on a global material infrastructure that includes satellites and scientific observational facilities, digital texts and memoranda, computer networks and atomic clocks, and the supply chains that support them. The machinery of chronometric governance is a sociotechnical infrastructure that in turn supports the greater undertakings of other large technical systems (Mayer and Acuto 2015). The present description of this infrastructure is necessarily brief but illustrates how extensive and pervasive in space *and time* is this network of chronometric endeavour.

Issues and Problems

The preceding narrative might easily give the impression that chronometric governance is a principally technocratic form of global governance in which intergovernmental organisations like the BIPM and ITU rub along in frictionless harmony with public scientific bodies, research institutions and specialist industries. This would be understandable, although misplaced. After all, the product of their mutually reinforcing activities is precisely that which they intend: a global time scale of utility to diverse practitioners, commerce, communications and which affords all the peoples of the world a firm reference point upon which to found their localised temporal practices, including, most visibly, national time zones and calendars. However, the mention of time zones reminds us that time is always exploitable, on account of its essentially constructed nature. Time zones are marked by abstract longitudinal lines that can be shifted to suit political and economic priorities, which also shift in time and space. North Korea's recent adoption of

'Pyongyang Time' – a snub to historical Japanese imperialism – is but the latest example of political manipulation of time zones (Harding 2015). In 2013, Samoa shifted the International Date Line eastward, bringing the country closer in time to its Australasian major trading partners and reversing a 19th-century decision that had taken it the other way (BBC 2011). In 2007, Venezuelan president Hugo Chavez shunted the national meridian westwards and national time thirty minutes back into a 'fractional time zone', ostensibly seeking 'a more fair distribution of the sunrise' (Reuters 2007). In 1949, Mao Zedong dispensed with the five time zones respected by the ousted nationalist government, unifying China into a single national time that persists today, despite its inconvenience to far-flung western provinces (Hassid and Watson 2014). The de facto adoption by western Uighurs of their own 'Urumqi Time' in resistance to 'Beijing Time', for example, continues not to be recognised by a Party-state concerned more with centralising control over 'national unity' than with respecting claims to provincial and ethnic autonomy (Schiavenza 2013).

States can elect to change national time, or to reject or embrace other measures like daylight-saving time, but they cannot easily opt out of the structures of chronometric governance. Technically, most countries are not signatories to many of the treaty instruments that govern time and other international standards – from the 1875 Metre Convention onwards – but they abide by those standards and implement changes when handed down by intergovernmental organisations like the BIPM and the International Organisation of Legal Metrology (OIML). Nor are the ordinary decisions of the ITU binding on its members unless implemented in national law. It is not legal sanctions that deter states from leaving but the benefits of 'membership' that make them stay (see Prakash and Potoski 2010). It is difficult to imagine any state wanting or being able to leave the system of global chronometric governance, given the imbrication of temporal standards with the sociomateriality of everyday life and national activity. This is not to condone the apparent homogenisation of global time through this system of global governance but it is to note the potency of its integrating logic.

As noted previously, the global temporal assemblage is more heterogeneous than perhaps meets the eye. Time is constructed but it is also contested, as a century-and-a-half of unresolved chronometric issues attests. Interest groups within the expert communities of metrologists and other scientists compete to produce temporal regimes most amenable to their interests. For many years, one such field of contention has been the issue of leap seconds. In 2015, leap seconds are, if not headline news, certainly a matter of public and therefore political concern. Since their formal inception in 1972, twenty-six leap seconds have been intercalated to align atomic time and astronomical time.[33] The most recent leap second was inserted at midnight on 30 June 2015. On that date, the stroke of midnight occurred twice, as clocks rolled over from 23:59:59 to 00:00:00 via 23:59:60, a phantom midnight that only exists on those occasions – about once every eighteen months – a leap second is required (BBC News 2015). There are ongoing discussions over whether to abolish leap seconds. Technically, this will depend on the redefinition (or not) of the word 'day'. Under a host of extant international agreements, a day is defined with reference to the rotation or location of the earth relative to the sun. To abolish leap seconds would require the decoupling of the day from the sun and linking it permanently to atomic time.

Abolitionists argue the link to solar time is no longer necessary on account of GPS and other satellite navigation systems that support location-dependent technologies like navigation and astronomy. They also make a strong case that leap seconds pose a threat to time-dependent technologies like the internet that were not designed to accommodate discontinuous time jumps. Serious disruptions to global communications networks and dependent technologies like banking and air traffic control have yet to occur when leap seconds have been added but the argument is that they might (Kamp 2011). Given the impossibility of re-engineering the internet, for instance, and the relative simplicity of abolishing leap seconds, this is a convincing argument to many. Proponents recognise that atomic and solar time will drift apart – albeit slowly – and argue that money is better spent preparing for occasional, longer time jumps (minutes or more) than on more frequent leap seconds introduced often at very short notice. Opponents argue that location-dependent tasks will require huge investments in software and hardware to allow for the continual divergence of solar and atomic time. Government consultations suggest that some members of the public recognise that leap seconds represent 'a symbolically important link with our past' (OPM Group 2014: 6). At the November 2015 World Radiocommunication Conference in Geneva, the ITU decided to retain leap seconds and therefore continue this link between atomic and solar time (ITU 2015). However, it also committed to further investigation into the feasibility of a new reference time scale, leaving open the possibility that there may be in future a final move towards a fully technological *chronos*.

Conclusion

This contribution to the emerging literature on time and temporality in International Relations introduces the idea that global time, principally Coordinated Universal Time (UTC), is the product of global governors operating in and through sociotechnical assemblages. Global chronometric

---

[33]   See, http://maia.usno.navy.mil/ser7/tai-utc.dat.

governance constructs UTC as a useful and unitary global time but this provisional analysis suggests that aspects of the global governance of time are contested. The example given here is of leap seconds but there are other topics that could be explored. The task of IR is to explore the modalities and topologies of global chronometric governance in order to better discern the activities and normative priorities of global governors and the constraints and opportunities provided by the structures of global chronometric governance. This is partly a matter of methods, as many of the actors involved – especially the ITU – operate under conditions approaching diplomatic secrecy and are somewhat opaque to outside observers. Careful empirical work is required to construct models of chronometric governance that can be tested against theories of global governance in IR.

Explorations of this type can also help to extend and strengthen our understanding of the mutually co-constitutive relations between time and politics. It is insufficient to identify the political nature of global time through its status as an object of global governance without also recognising the temporal nature of global politics. World politics is, as Kimberly Hutchings asserts, 'a shifting and unpredictable conjunction of times' (Hutchings 2008: 176), unified only by the analytical lens chosen to examine them. There is no pristine ontological *chronos* but rather a heterotemporal assemblage of times and temporalities that is both the cause and outcome of political behaviours seeking to maximise self-interest and effect social change. Time is not a 'background condition' of global life but is 'socially constructed and therefore amenable to manipulation by human agency' (Porter and Stockdale 2015: 12). As this discussion of UTC aims to have shown, this manipulation extends to the scientific and technocratic construction of global time, derived from modern physics as much as ancient astronomy.

The purpose of the present enquiry is to provide empirical support for the proposed heterotemporality of global time, an inherently political activity about a richly textured political 'object'. If we accept its political nature, how else do we open up this particular 'black box' to scrutiny and critique? What avenues of contestation are available, if indeed such a thing is necessary? What does it mean in the present discussion to cast off from our nearest star and look inwards to the oscillations of an invisible atom? Does it matter if pragmatism trumps philosophy? These are all questions that might be addressed through further exploration of the global governance of time.

References

Audoin, C. and Guinot, B. (2000) *The Measurement of Time: Time, Frequency and the Atomic Clock.* Cambridge: Cambridge University Press.

Avant, D., Finnemore, M. and Sell, S.K.S. (2010) 'Who governs the globe?' In: Avant, D. D., Finnemore M. and Sell, S.K. (eds) *Who Governs the Globe?* New York: Cambridge University Press. 1-31.

Barbour, J. (1999) *The End of Time: The Next Revolution in Our Understanding of the Universe.* Oxford: Oxford University Press.

BBC News (2011) 'How does a Country Change its Time Zone?' 10 May. Available online at: http://www.bbc.co.uk/news/world-13334229.

BBC News (2015) '"Leap second" Added for First Time in Three Years.' 1 July. Available online at: http://www.bbc.co.uk/news/science-environment-33313347.

Beard, R. L. (2011) 'Role of the ITU-R in time scale definition and dissemination.' *Metrologia* 48(4): S125-S131.

Berenskoetter, F. (2011) 'Reclaiming the vision thing: Constructivists as students of the future.' *International Studies Quarterly* 55(3): 647-68.

Bureau International des Poids et Mesures (BIPM) (2006) *The International System of Units (SI)*. 8th. edn. Paris: Organisation Intergouvernementale de la Convention du Mètre.

Harding, L. (2015) 'North Korea to Turn Clocks Back by 30 minutes to Establish Own Time Zone.' *The Guardian*. 7 August. Available online at: http://www.theguardian.com/world/2015/aug/07/north-korea-to-turn-clocks-back-by-30-minutes-to-establish-own-time-zone.

Hassid, J. and Watson, B.C. (2014) 'State of mind: Power, Time Zones and Symbolic State Centralisation.' *Time & Society* 23(2): 167-94.

Hofferberth, M. (2015) 'Mapping the Meanings of Global Governance: A Conceptual Reconstruction of a Floating Signifier.' *Millennium: Journal of International Studies* 43(2): 598-617.

Hom, A. R. (2010) 'Hegemonic Metronome: The Ascendancy of Western Standard Time.' *Review of International Studies* 36(4): 1145-70.

Hom, A. R. (2012) 'Two Regimes of Time.' *E-International Relations*. Available online at: http://www.e-ir.info/2012/12/24/two-regimes-of-time/.

Hom, A. and Steele, B.J. (2010) 'Open Horizons: The Temporal Visions of Reflexive Realism.' *International Studies Review* 12(2): 271-300.

Hutchings, K. (2007) 'Happy Anniversary! Time and Critique in International Relations Theory.' *Review of International Studies* 33, supplement S1: 71-89.

Hutchings, K. (2008) *Time and World Politics: Thinking the Present.* Manchester: Manchester University Press.

International Telecommunication Union (ITU) (2002) 'Recommendation ITU-R TF.460-6. Standard-frequency and Time-signal Emissions.' Available online at: https://www.itu.int/rec/R-REC-TF.460/en.

International Telecommunication Union (ITU) (2015), 'Coordinated Universal Time (UTC) to retain "leap second"', press release, 19 November.

Kamp, P. (2011) 'The One-Second War.' *Communications of the ACM* 54(5): 45-8.

Karol, P.J. (2014) 'Weighing the Kilogram.' *American Scientist* 102(6): 426-9.

Kennelly, A. E. (1935) 'Adoption of the Meter-kilogram-mass-second (MKS) Absolute System of Practical Units by the International Electrotechnical Commission (IEC), Bruxelles, June, 1935.' *Proceedings of the National Academy of Sciences of the United States of America* 21(10): 579-83.

Klinke, I. (2013) 'Chronopolitics: a Conceptual matrix.' *Progress in Human Geography* 37(5): 673-90.

Kütting, G. (2001) 'Back to the future: Time, the environment and IR theory.' *Global Society* 15(4): 345-360.

McIntosh, C. (2015) 'Theory across Time: the privileging of Time-less theory in International Relations.' *International Theory* 7(3): 464-500.

Mayer, M. and Acuto, M. (2015) 'The Global Governance of Large Technical Systems.' *Millennium: Journal of International Studies* 43(2): 660-83.

Mazower, M. (2012) *Governing the World: A History of an Idea.* London: Penguin.

Nelson, R.A., McCarthy, D.D., Malys, S., Levine, J., Guinot, B., Fliegel, H.F., Beard R.L. and Bartholomew, T.R. (2001) 'The Leap Second: Its History and Possible Future.' *Metrologia* 38(6): 509-29.

OPM Group (2014) *Leap Seconds Dialogue: Final Report.* London: OPM Group. Available online at: http://leapseconds.co.uk/wp-content/uploads/Leap-seconds-dialogue-final-report.pdf.

Palmer, A.W. (2002) 'Negotiation and resistance in global networks: the 1884 International Meridian Conference.' *Mass Communication & Society* 5(1): 7-24.

Porter, T. and Stockdale, L.P.D. (2015) 'The Strategic Manipulation of Transnational Temporalities.' *Globalisations* DOI: 10.1080/14747731.2015.1056497.

Prakash, A. and Potoski, M. (2010) 'The International Organisation for Standardisation as a Global Governor: a Club Theory Perspective.' In: Avant, D., Finnemore, M. and Sell, S. (eds) *Who Governs the Globe?* New York: Cambridge University Press. 72-101.

Reuters (2007) 'Country to change time zone by 30 minutes.' 24 August. Available online at: http://www.reuters.com/article/2007/08/24/us-venezuela-time-idUSN2328980320070824.

Rosenau, J.N. and Czempiel, E. (eds) (1992) *Governance without Government: Order and Change in World Politics.* Cambridge: Cambridge University Press.

Schiavenza, M. (2013) 'China Has Only one Time Zone—and That's a Problem.' *The Atlantic.* Available online at: http://www.theatlantic.com/china/archive/2013/11/china-only-has-one-time-zone-and-thats-a-problem/281136/.

Stevens, T. (2016) *Cyber Security and the Politics of Time.* Cambridge: Cambridge University Press.

Zumbansen, P. (2012) 'Governance: an interdisciplinary perspective.' In: Levi-Faur, D. (ed.) *The Oxford Handbook of Governance* (Oxford: Oxford University Press). 83-96.

5

Calendar Time, Cultural Sensibilities, and Strategies of Persuasion

KEVIN K. BIRTH

QUEENS COLLEGE, CUNY, USA

In considering the relationship of time, globalisation, and international relations, time cannot be viewed as an abstract, uniform yardstick that merely provides a chronological organisation of events. Instead, there is a close relationship between cultural ideas of time and political sensibilities. There are many elements to this connection that one could explore, but here I shall limit myself to the relationship between calendars, holidays, and styles of persuasion. To do this, I shall discuss the tensions generated from the global distribution of Western timekeeping; describe the values latent in the secular holidays of the United States, the United Kingdom, and People's Republic of China; and show the relationship between the values celebrated in calendars and the rhetorical strategies used by these nations in two very different international debates: climate change and the leap second.

There are relatively few political debates about concepts of time. Yet, there are few ideas that pervade almost everything to the same extent as temporal ideas. The value of money and securities is influenced by trading protocols in which the sequence of trades is documented by means of synchronised timestamps. Getting directions from a Global Positioning System (GPS) involves relying on a suite of satellites that are basically orbiting clocks transmitting time signals. National security relies on the analysis of big data sets in which precise timestamps are crucial to understanding the sequence and pattern of events.

Anthropologists are interested in how concepts of time are related to the exercising of power and the structuring of political and social action (Rutz 1992; Greenhouse 1996). Public consciousness of this fact is probably not as great now as it was in the past when heads of state controlled calendars and state defined holidays and thereby controlled the rhythms and celebrations in people's lives (see Stern 2012). Until the seventeenth century, many European chronologies referred to the reigns of monarchs (Wilcox 1987), and many civilisations—China, the Maya, Rome—involved explicit links between those who defined time and the ruling elite.

The currently dominant system of clock and calendar time emerged in association with European imperialism, and it gained its current distribution as a result of colonialism and global trade (Bartky 2007; Birth 2012; Quinn 2012). Capitalism has been shaped by the logic of fixed terms and timed transactions and wages have been defined in terms of units of time rather than directly in terms of the ideal timing for most the most productive work. The time grid, established by the Gregorian calendar and clock time, is also a tool in bookkeeping—time becomes represented as containers to be filled and/or audited. Modern governance would be quite different without the relationships among global units of time, management, and bookkeeping, particularly since these units of time are used as a means of control and discipline. In late June 2015, there was a confrontation between the International Monetary Fund (IMF) and Greece over a debt repayment due 1 July 2015. The head of the IMF, Christine Lagarde, stated unequivocally that the 1 July deadline was non-negotiable. There is no solar, lunar, astronomical, or biological cycle that defines 1 July, however. Its reality is due to cultural convention, but that cultural convention is a political tool.

The European Cultural Logic behind the Global Time System

This time system now reaches into the structure and design of computer systems. Everything a computer does receives a timestamp, and commands are often executed in a sequence based on these timestamps. Timestamps are globally synchronised to Coordinated Universal Time (UTC)—a timescale maintained by the International Bureau of Weights and Measures (BIPM). The BIPM determines UTC using measurements received from atomic clocks distributed across the globe. As a result, the definition and distribution of time has reached unprecedented levels of precision and accuracy.

It must be pointed out, however, that the emphasis on hyper precision is the result of a cultural, not a global, choice. Unlike timekeeping in other parts of the world, European timekeeping came to emphasise the use of uniform units of duration divorced from observable astronomical cycles to represent time (Birth 2012). This temporal logic is embedded in European-style clocks (Borst

1993; Dohrn-van Rossum 1996; Landes 1983), the default form of clock found in the world today. Other cultures have had other logics. Jewish zmanim and Christian canonical hours are defined in terms of seasonally variable hours that served as points in time rather than as set durations. Edo period Japanese clocks divided the day into six daylight and six nighttime seasonally variable hours, and they created clocks to represent these seasonal variations. Chinese timekeeping and Hindu jyotish (astrology) are anchored to the interaction of celestial cycles. European timekeeping privileges uniform oscillations for defining time over the irregular rotational behaviour of Earth.

The Global Distribution of Western Time Concepts

The current use of Western time standards across the globe was not the result of an election or national leaders signing a treaty. It is a Western scientific preference riding on the coattails of colonialism (see Barak 2013). The fixed and uniform nature of time metrology makes it appear as if it is apolitical, yet decisions such as where to place the prime meridian for timekeeping purposes were hotly contested (see Barrows 2011), and there continue to be global debates about the definition of time, such as the leap second debate that is discussed later.

The far reach of Western time technology and Western time logics creates peculiarities in the world. GPS was developed as a United States military technology. It consists of clocks mounted on satellites that emit time signals. GPS is now critical to the functioning of all sorts of software applications, including those used to indicate Islamic prayer times and the direction of Mecca from any position on the globe on mobile devices. Somewhere, a member of ISIS is using a Muslim prayer application for this purpose—his/her devotion is achievable with great precision in timing and orientation because of the assistance of the US military's navigation satellites. In effect, even the sworn enemies of the US employ US military technology to know time and place.

Adaptation versus Adoption

While GPS and precision time technology might be viewed as being of great benefit to all of humanity, GPS time represents the imposed Western time. Much of the world has figured out how to adapt itself to Western clock and calendar time, but for many, it is an adaptation, not an adoption. In some parts of the world, the Gregorian calendar is associated with a religion (Christianity) and colonising nations towards which there is antipathy.

To get a sense of adaptation rather than adoption, try to place a special holiday in a different calendar. Here, I shall use the example of Christmas. Its date is determined by the Gregorian calendar—it is always December 25. But if the Gregorian calendar was not dominant, then Christmas would appear differently in calendars.

For the sake of imagination, let us conceive that the dominant calendar was the Islamic calendar, and that Christmas would need to be determined first using the Gregorian calendar, and then placed within the Islamic calendar. This situation is already what many people throughout the world have to do— indeed, there are more Hindus, Chinese, Jews, and Muslims adapting their holy days to a Gregorian rubric than Westerners for whom the Gregorian calendar represents their heritage. To place Christmas in a non-Gregorian calendar would involve using algorithms such as those in Dershowtiz and Reingold's *Calendrical Calculations* (1997). Such algorithms are now commonplace in applications that take computer dates and times that are in a Gregorian and UTC format and convert them into religiously significant times for Jews, Hindus, and Muslims.

Placing Christmas in the Islamic calendar faces an additional complication however—there is no single, globally agreed-upon Islamic calendar. This emerges from a debate about whether the beginning of the lunar month can be predicted or can only be announced after the new moon is observed. It is also complicated by the debate over whether it is the local sighting of the moon that is important, or the appearance of the moon in Mecca. Important Muslim holy days, like the month of Ramadan, can begin on different days for different Muslims depending on which calendrical authorities they follow. As a result, any attempt to relate Christmas to an Islamic calendar also involves a *political* choice of which Islamic calendrical method to use—a choice that brings with it an implicit privileging of one source of Islamic calendrical authority over others.

Once the problem of placing Christmas in the Islamic calendar has been resolved, there is another question of what to do when Christmas occurs during Ramadan, the Muslim month of fasting. This happens because the Islamic year consists of 12 lunar months and is about 11-12 days shorter than the solar year that forms the basis of the Gregorian calendar. As a result, holidays in the Gregorian and Islamic calendars drift in relation to one another. The last time Christmas was held during Ramadan would have been the period from approximately 1419-1421 in the Islamic chronology (1998-2000 in the Christian chronology).

Adaptation and Conflict

It does not take much imagination, then, to realise the extent to which calendrical differences can highlight differences in a globalised world. The domination of the Gregorian calendar and Western clock time allows those who know no other calendar or clock to avoid seeing the calendrical conflicts. Indeed, Western clock and calendar time come to be viewed as natural and beyond question. With increased contact between people of different cultural traditions, conflicts can occur along with entrenched misunderstandings.

For instance, a few years ago, a temporary instructor at my institution complained bitterly about the Jewish High Holy Days. She wanted to give an exam on one of those days and did not want to excuse Jewish students from the exam. Her argument was that since the university was a secular institution it used a secular calendar and should not recognise religious holidays. Because such clashes have been a problem in the past, the university has adopted a policy that such students are entitled to take the exam on another day. The commonly held Gregorian calendar does not eliminate difference—it merely elides it. Conflicts occur, sometimes with unhappy results. While my department forced the instructor to comply with university guidelines, this did not make the instructor very happy, and she continued to express resentment towards the Jewish students in question.

This type of situation is not new. While it seems to have little bearing on modern globalisation, I shall use the example of medieval European Judaism to briefly explore how antagonism gets intermingled with calendars. The choice of this example is that the time that has passed since this case makes it easier to discuss than some current events, but the lessons that can be learned are still relevant. Many medieval European Jewish populations recorded the days of Christian saints in their calendars. At a functional level, this makes sense. Stern (n.d.) points out that markets and fairs were tied to Christian saints' days and that this information would have been useful to Jewish traders. He also suggests that dating official documents would have also benefited from knowledge of saints' days. After all, many of the commercial cycles were directly tied to the liturgical calendar, such as the British banking cycle that defined the quarterly periods with reference to Trinity Sunday and Michaelmas.

While Jewish calendars made reference to saints' days, this was not a sign of collegial ecumenicalism between Jews and Christians in the Middle Ages. Anti-Semitism and persecution of Jews by Christians in the Middle Ages are well documented, and Jewish sentiments about Christianity are expressed in their calendars. Stern (n.d.) has described the anti-Christian rhetoric found in them. For instance, in several calendars, the Hebrew spelling of the word 'saints' can be read *qedeshim*, which means 'prostitutes' (Stern n.d., 14, 28). In another example, Stern studies a North French calendar from 1278 which begins with the statement 'these are the months of the non-Jews and their abominations' (Stern n.d., 4).

Holidays and Cultural Sentiments

The annual cycle of holidays reveals cultural sentiments. One can look at the secular holidays adopted by different nations and quickly ascertain different emphases on the sorts of events and people commemorated (Callahan 2006). One can also look at school curricula and often see these emphases reflected in education. But what is most disconcerting is to look at national policies and see how the logics that guide the holidays and the schooling shape policy arguments. In thinking about time and politics, it is not merely a matter of how Western time reckoning has been imposed on the globe, but how the sentiments reflected in the holidays shape how global events and conflicts are discussed. There is a striking correspondence between the values embodied in holidays, and the types of arguments and evidence policy makers use in staking out their positions. In drawing this connection, my point is less to identify policy differences than to highlight differences in the sorts of arguments and evidence thought to be persuasive, but which might, in fact, be unpersuasive across cultural differences.

The national secular holidays in the United States are: Martin Luther King Jr. Day, President's Day, Memorial Day, Independence Day, Labor Day, Columbus Day, Veteran's Day, and Christmas. New Year's is left off of this list because of its religious origins within the British Empire and its former colonies—it was the Feast of the Circumcision whereas 25 March was the beginning of the new year until 1752. Among the secular American holidays, four directly refer to war: Presidents' Day, Memorial Day, Independence Day, and Veterans' Day. Two refer to martyrs for social causes: Martin Luther King Jr. Day and Labor Day. Christmas's history is actually tied to war since it was moved to November by Abraham Lincoln during the Civil War in order to celebrate what he perceived as the turning point in the conflict. Moreover, many people, particularly Native Americans, associate Christmas and Columbus Day with the killing of America's indigenous population. In effect, every single secular holiday in the United States has an association with a war, an armed conflict, or people being killed.

In contrast, the United Kingdom's secular holidays are known as bank holidays. These holidays include Spring Bank Holiday, Late Summer Bank Holiday, and Boxing Day. The Spring Bank Holiday replaced the religious holiday Whit Monday. Whit Monday's date varied because its timing was tied to Easter. Now, The Spring Bank Holiday falls on the last Monday in May regardless of when Easter occurs. Boxing Day does not celebrate banking, but it does emphasise social stratification. If one looks for themes in British secular holidays, they are banking and social stratification.

In the People's Republic of China, secular holidays come in two varieties. One set of holidays celebrates the traditional holidays of the Chinese calendar. As a result, these holidays shift from one year to the next in the Gregorian calendar. Several holidays within the traditional calendar have clear religious connotations, such as Chinese New Year's Day and the Qingming Festival. Others, such as the Duanwu Festival (Dragon Boat Festival) and the Mid-Autumn Festival emphasise astronomical events: the summer sun and the harvest moon, respectively. As such, they are associated with food, feasting and agriculture. China also has three holidays tied to the Gregorian calendar: New Year's Day, Labour Day, and National Day. National Day not only celebrates Chinese Nationalism, but Chinese industry and economic development, as well. There is a twofold theme in the holidays of the People's Republic of China, then: traditional time reckoning, and an emphasis on labour, agriculture, industry, and economic development.

Holidays indicate important cultural sentiments. The sentiments reflected in national holidays also manifest in styles of persuasion—an uncanny and sometimes unsettling relationship between how time is structured and celebrated and how nations participate in global debates. To demonstrate this, I shall focus on two sets of governmental statements of policy positions: climate change and the leap second.

Time, Cultural Sentiments, and Climate Change Rhetoric

Of these two policy debates, the issue of climate change is well known. While one normally thinks of this debate as framed in scientific terms, one can see cultural differences in the approach to the climate change issue in the key documents released by the United States, the United Kingdom, and the People's Republic of China. All three nations are very concerned about climate change, but the ways in which the concerns are stated is quite different from one another.

The key document in stating the US position is the report *Climate Change Impacts in the United States* (Melillo et al., 2014). This is an 841-page report in which something is described as being killed or dying approximately every 10 pages. In addition, whereas most of the report addresses how climate change will affect food production, the conclusion emphasises the need to understand how climate change will affect the military. The fascination with death and the military is consistent with the sentiments cultivated by American holidays. Moreover, the second most important document about climate change is the 2014 US Department of Defense's *Climate Change Adaptation Roadmap* (2014). In fact, President Obama often cites this document generated by the military rather than the more thorough *Climate Change Impacts*. American holidays focus on death and war; the United States government's articulation of climate change policy focuses on death and war.

In contrast, Great Britain's key document, *The National Adaptation Programme* (HM Government, 2013) emphasises investments. As the ministerial foreword to this report states, 'Britain has a long history of overcoming the challenges that our famously changeable weather poses and harnessing our natural resources to support growth. New investments and innovation in both the private and public sectors continue this tradition today' (2013, 1). Banking or investment is discussed on average once every two pages in this document. In contrast to the United States' documents, the British document does not mention anything dying or being killed, and the military is only mentioned once.

The key document from the People's Republic of China is *China's Policies and Actions for Addressing Climate Change* (People's Republic of China, The National Development and Reform Commission 2013). In this report, industry or industrialisation is mentioned on every single page. Like the British document, there is no killing or dying, and there is no mention of the military, either.

In effect, the ways of thinking about climate change in public policy reports reflects the cultural sentiments found in the national holidays: the US emphasises the military, death, and dying; the UK emphasises banking and investment; the People's Republic of China emphasises industry.

This is just one example of how the structure of time throughout the year subtly shapes policy rhetoric if not policy initiatives. This pattern also holds true for the leap second debate—a debate almost entirely unrelated to climate change.

Time, Cultural Sentiments, and the Leap Second Debate

The term 'leap second' refers to a solution to the problem of reconciling a timescale defined by means of atomic timekeepers with the Earth's rotation (see Stevens 2015).

A consequence of the leap second policy is that the majority of the world's population, concentrated as it is in east Asia, has to put up with the leap second being implemented at the beginning of the business day—the time equivalent to midnight at the prime meridian. This is an unconscious consequence of the explicit Eurocentric assumptions that guided the original definition of Universal Time in 1884, such as the following statement by Great Britain's Astronomer Royal W. H. M. Christie in 1886:

> The advantage of making the world day coincide with the Greenwich civil day is that the change of date at the commencement of a new day falls in the hours of the night throughout Europe, Africa, and Asia, and that it does not occur in the ordinary office hours (10 a.m. to 4 p.m.) in any important country except New Zealand' (1886, 523).

This logic has, over a century later, resulted in a European minority of the world's population sleeping through any disruptions caused by the leap second while others have to contend with the leap second coinciding with the beginning of morning business cycles.

Eventually, problems with the leap second caught the attention of the West. In the 1990s, as networked computing became important and reliant on precision timing, some began pointing out problems generated by the leap second. Timestamps are a means of maintaining computer security across networks. If a computer receives a bit of code from another computer with a timestamp outside of the expected range, then it is deemed a security threat—it is considered as either a fraudulent timestamp or, if the timestamp indicates an unexpected delay in the signal's travels then, it indicates the possibility of somebody inserting malicious code somewhere along the way, thereby causing the delay. So, unexpected timestamps tend to result in computers rejecting the associated information or commands. If such rejection occurs throughout a networked system, it can cause system-wide problems and eventually cause servers to crash.

Since leap seconds are not predictable the way that leap years are, they cannot be written into operating systems. Instead, when a leap second is announced, software patches need to be developed to ensure that a computer handles the leap second correctly. If a systems administrator does not handle the leap second correctly, then the system might go down, as when Qantas Airway's reservation system crashed after the 2012 leap second.

The leap second policy is decided by the Radiocommunication Sector of the

International Telecommunications Union (ITU-R)—an agency of the United Nations. In 2001, the ITU-R called for research on the leap second question, and there has been a policy debate ever since then over whether or not to keep the leap second. The United States, the United Kingdom, and the People's Republic of China have adopted differing positions and different argument style in advocating their positions.

Given the militaristic sentiments of the United States, it is no surprise that the leap second debate emerged in association with an American military technology as it was made available for civilian use: GPS (McCarthy and Klepczynski 1999). As the debate has unfolded, those most vocally involved have been American astronomers, computer programmers, time metrologists associated with the United States Naval Observatory or the United States National Institute of Standards and Technology, and those who work for the Department of Defense or military contractors. Consistent with American cultural logic emphasising death, one of the most common arguments for eliminating the leap second policy is that eventually a leap second will cause a problem such that 'planes will crash and people will die' (Allen 2013; Sobel 2013; Wolman 2013).

The British position is to keep the leap second. This position is closely linked to the sense of the prime meridian in Greenwich being a part of British heritage. Eliminating the leap second would decouple timekeeping from the prime meridian, and as the former minister of Science and Universities complained, eventually the meridian for midnight would be over the United States (Swinford 2014). Truthfully, it probably would not even reach Big Ben in that minister's lifetime, however. In 2014, the United Kingdom hired a research firm to conduct a study to learn how the British public felt about the issue (Silver et al. 2014). Since most members of the public knew nothing of the issue, much of the study involved educating focus groups about the leap second, and not surprisingly, such efforts at education contained subtle biases towards the British position to keep the policy. That said, the conclusion of the study was that unless there were strong technological or financial cases made for getting rid of the leap second, most members of the British public would not support doing so (Silver et al 2014, 52).

Once again, while the American position is tied to the military and people dying, the British position highlights banking and finance. The Chinese position has been evolving. Initially, representatives of the People's Republic of China's argued that since the solar day was important to the Chinese people, decoupling the global timescale from the Earth's rotation would be disruptive (Han 2013). In 2015, at a conference in Australia, the Chinese representative, Dr. Han Chunhao, offered a different argument. He documented traditional Chinese timekeeping practices and made the argument that those practices were just as scientific as the timekeeping practices of the West. Therefore, while he supports eliminating the leap second policy, he advocates creating a means of disseminating Earth rotational information so that traditional Chinese timekeeping could be maintained, and he suggests that the global navigation satellite systems be used for this purpose. Referring back to the People's Republic of China's secular holidays—most of them are tied to the traditional Chinese calendar. The Chinese leap second position, like the US and UK positions, reflects the sentiments expressed in public holidays. Returning to the points made earlier, it also reflects the Chinese consciousness of being forced to adapt to Western time-keeping, and a desire for the West to reciprocally adapt to support Chinese time-keeping.

Conclusion

The relationship of time, politics, and globalisation involves the interaction of the global imposition of a Western timescale, local ideas of timekeeping, and how cycles of holidays shape sentiments and approaches to political challenges. Most of these issues lurk beneath the surface of consciousness—overshadowed by ideologies such as the world being flat (Freidman 2005), or time being 'natural'. Even in debates about time policies, such as the current leap second debate, the sentiments reinforced, if not shaped, by national holidays still influence the positions and arguments that representatives from different nations make. Consistent with the history of the global time system since 1884, when Universal Time and the prime meridian were determined, a few European nations and the United States have an influence over time policies that is disproportionate to their percentage of the world's population. The leap second debate is largely a debate between Western time metrologists that the world has been asked to adjudicate through the ITU-R.

The discussion of climate change indicates that the way in which temporalities shape sentiments goes far beyond time policies, however, but affects other, if not all, issues of global importance. The similarity of arguments about leap seconds and climate change suggests cultural inclinations creeping into scientific and policy debates in ways that are rarely, if ever, acknowledged. Whether it is the sentiments cultivated by different calendars and holidays, or the conflicts generated when different temporalities collide, time exerts a subtle and pervasive force on political activity and negotiation—a force that is rarely recognised.

* The author would like to thank Sacha Stern for sharing his research on medieval Jewish calendars, the participants in the Utopias, Futures and

Temporalities symposium in Bristol in May 2015 for their feedback on some of the ideas expressed here, and the anonymous reviewers for their constructive comments on this chapter.

References

Allan, S. L. (2013) 'Planes Will Crash! Things that Leap Seconds Didn't, and Did, Cause.' In: Seago, J. H., Seaman, R. L., Seidelmann, P. K., and Allen, S. L. (eds) *Requirements for UTC and Civil Timekeeping on Earth*. American Astronautical Society, Science and Technology Series vol.115. San Diego: Univelt.

Barak, O. (2013) *On Time*. Berkeley: University of California Press.

Barrows, A. (2011) *The Cosmic Time of Empire.* Berkeley: University of California Press.

Bartky, I. (2007) *One Time Fits All.* Stanford: Stanford University Press.

Birth, K. K. (2012) *Objects of Time.* New York: Palgrave Macmillan.

Borst, A. (1993) *The Ordering of Time.* Chicago: University of Chicago Press.

Callahan, W. A. (2006) 'War, Shame, and Time: Pastoral Governance and Naitonal identity in England and America.' *International Studies Quarterly* 50: 395-419.

Christie, W. H. M. (1886) Universal or World Time. *Nature*. April 1: 521-523.

Derschowitz, N. and Reingold, E. (1997) *Calendrical Calculations.* Cambridge: Cambridge University Press.

Dohrn Van Rossum, G. (1996) *History of the Hour*. Chicago: University of Chicago Press.

Friedman, T. L. (2005) *The World is Flat: A Brief History of the Twenty-First Century*. New York: Farrar, Straus and Giroux.

Greenhouse, Carol. (1996) *A Moment's Notice: Time Politics Across Cultures*. Ithaca: Cornell University Press.

Han Chunhao (2013) 'Space Odyssey: Time-scales and Global Navigation Satellite Systems.' *ITU News*. Number 7 (September).

HM Government (2013) *The National Adaptation Programme: Making the Country Resilient to a Changing Climate.* London: The Stationery Office.

Landes, D. (1983) *Revolution in Time*. Cambridge, MA: Belknap.

McCarthy, D.D. and Klepczynski, W. J. (1999) 'GPS and Leap Seconds: Time to Change?' *GPS World*. November: 50-54.

Melillo, J. M., Terese (T.C.) R. and Yohe, G.W. (eds) (2014) *Climate Change Impacts in the United States: The Third National Climate Assessment*. Washington, D. C.: Global Change Research Program.

Quinn, T. (2012) *From Artefacts to Atoms*. Oxford: Oxford University Press.

People's Republc of China. National Development and Reform Commission (2013) *China's Policies and Actions for Addressing Climate Change*. People's Republic of China.

Rutz, H. J. (1992). 'Introduction: The Idea of a Politics of Time.' In: Rutz, H. J. (ed.) *The Politics of Time*. American Ethnological Society Monograph Series, Number 4. Washington, D. C.: American Anthropological Association.

Sobel, D. (2013) 'The Wait of the World.' *Aeon*. July 11. Available online at: http://aeon.co/magazine/science/what-if-clock-time-no-longer-tracked-the-sun/

Silver, K., Wild, M. and Bedford, T. (2014) *Leap Seconds Dialogue: Public Dialogue Report*. London: OPM.

Stern, S. (2012) *Calendars in Antiquity: Empires, States, and Societies*. Oxford: Oxford University Press.
Stern, S. (n.d.) Christian Calendars in Hebrew in Jewish Medieval Manuscripts.

Stevens, T. (2016) 'Governing the Time of the World' in Hom et al, *Time Temporality and Global Politics*, Bristol: E-International Relations.

Swinford, S. (2014) 'Greenwich Mean Time could drift to the US, minister warns.' *Daily Telegraph.* May 14. Available online at: http://www.telegraph. co.uk/news/politics/10831974/Greenwich-Mean-Time-could-drift-to-the-US-minister-warns.html

US Department of Defense (2014) *Climate Change Adaptation Roadmap.* Alexandria, VA: Office of the Deputy Under Secretary of Defense for Installations and Environment. Science and Technology Directorate.

Wilcox, D. J. (1987) *The Measure of Time Past.* Chicago: University of Chicago Press.

Wolman, D. (2015) 'Just a Second: The Leap Second May Be a Ticking Time Bomb.' *Slate.* June 10. Available online at: http://www.slate.com/articles/ health_and_science/science/2015/06/next_leap_second_june_30_dangers_ to_software_military_gps_banking_air_traffic.single.html

6

Analogue Time, Analogue People and the Digital Eclipsing of Modern Political Time

ROBERT HASSAN
UNIVERSITY OF MELBOURNE, AUSTRALIA

> ...*political time is out of synch with the temporalities, rhythms,
> and pace governing economy and culture.*
> Sheldon Wolin (1997:4)

It has been said, perhaps apocryphally, that in 1972 when asked by a Western reporter of his assessment of the 1789 French Revolution, Zhou Enlai, the Chinese prime minister and long-time ally of Chairman Mao, replied that 'It's too soon to tell'. This is one of those tales where it does not matter if it is true or not—it's amusing and thought-provoking at the same time. There's value in that Cho's alleged words may be read as both glib and serious; but what is interesting is that there's a latency as well as a potential *revealing* that stems from the glibness, and this serves to accentuate the seriousness of the point being made.

So what *is* the point? Well, at the risk of putting more words into Cho's mouth, one could have him say that however suddenly a revolution may spring up, to achieve its aims it will necessarily be a *long-run affair*. The passing of a couple of centuries would be the *least time* it would take before an event so historically profound as the French Revolution—by many a hoped-for precursor of the utter transformation of the world by putting Enlightenment thinking into practice—could even begin to look some way assessable, either positively or negatively. When we speak of time in this broad and general way, though, as we tend to do, we miss something very important about the *nature of time* and about what might be called *political time*. And when we think about the nature of political time in the present era, which we will do presently, then disturbing realisations about the future of the institutional political process—politics at both the local and global level—become noticeable. In short, the forms of politics that shaped our modern world were themselves evolved and shaped through a particular relation to time, analogue time. However, this time, and these political processes are being eclipsed and rendered less effective by a new temporal context, a digital context based upon computer networks. The internet and its myriad appurtenances generate and sustain a post-modern *network time* where the long-run political affair characteristic of the modern—be it the evolution of democratic processes between nation-states (Fukuyama, 2011: 245-458), or slow-burning gradualism within states (Havel, 1990: 115)—becomes less sustainable.

Network time suggests an accelerating time, an idea that political theorists such as Kimberley Hutchings (2008:17) and Bill Scheuerman (2004:26-71) argue, is becoming increasingly relevant to how we understand the political process. Some, indeed, such as William E Connolly (2002: 140-176) see social acceleration as potentially a positive thing for democracy. One can certainly agree with thinkers such as these that time is transforming how politics is enacted. However, much of this theorising takes place at a certain level: for example, Scheuerman's otherwise excellent book on liberal democracy and the acceleration of time tells us nothing about why our temporal relationships have been transformed. Likewise, Connolly's nuanced and flexible approach, says a good deal about *speed*, but little about the human experience of time—still less how it may be mediated through communications technology. What this chapter will do is locate the theorisation at what I understand to be the central issue in our new experience of time and politics—which is the relationship between analogue and digital processes, and how the individual and the collective and hence the political, are situated within this new *technological* context. Before we come to that, we need first to place time, technology and politics into a wider frame.

Time and the Materialisation of Political Ideas

The rhythms of time for the revolutionary process that began in 1789 were generated by the communication networks that allowed political ideas to disseminate, take root and grow. To understand how this worked, French philosopher Régis Debray laid stress upon what he termed 'the material forms and processes through which [political] ideas were transmitted—the communication networks that enable thought to have social existence' (2007: 5). In other words, to know how ideas materialise from abstract thought to become social reality, we need to know the nature of the communication technologies that relay them. More succinctly: when we speak of political ideas, the medium is a major part of the message.

The epoch that stretched between the Storming of the Bastille and Zhou Enlai's alleged aside to a reporter, Debray labelled the 'graphosphere', or the age of writing. Historian Dena Goodman (1994) referred to the early phase of this era as the 'republic of letters', where major philosophers and politicians and thinkers created their own networks of correspondence that served to materialise their thought and transform their world as a consequence; and more broadly the cultural theorist Benedict Anderson (1982) saw the period as helping to generate a *modern* culture that was based upon the writing, reading and circulation of the printed word to a growingly populous and literate public sphere.

The temporality of the graphosphere was given precision and predictability in its rhythm by the clock, and the clock in its turn gradually regulated and disciplined society as a whole in its business, its industry and its cultures. It created what Paul Virilio (2012:23) termed 'the administration of duration'. Of course politics and the political process were not exempt from this. And so as Western industrialising societies rose and developed to this clock-based rhythmicity, a 'clock-time consciousness' as social geographer Nigel Thrift (1996: 174) termed it, gave shape and context to what he saw as 'the cognitive framework within which political knowledge is interpreted' (p.167).

All this rhythmic coordination was developed and interacted at a pace that was, broadly speaking, a human pace; one that humans more or less could cope with, react to, and reflect within. As noted above, with technological progress, the modern world seemed (has it not always?) to be accelerating— to be slipping away from human control, and given over to the power of machines and industrialisation. Trains, telegraphs, motorcars and airplanes all contributed to the growing rapidity of the world as an increasingly sophisticated communicative space. However, the graphosphere was also an *analogue sphere*. By analogue I mean that its technologies have equivalence with nature and/or correspond with nature in some recognisable way. Indeed, writing – that fundamental communication technology – is itself analogue in both these senses: the written word has an equivalence with the spoken word, and early pictographic writing corresponded to the natural world in its depiction of people, animals, tools and so on. More recent transformative technologies are recognisably analogue also in that the motorcar has its correspondence in the horse, and the airplane finds its equivalent in nature in the bird. The centrally important temporal technology of the clock of course is analogue too in that it roughly corresponds with the rotation of the Earth and its revolutions around the sun.

As Silvia Estévez puts it, for most of human history technologies can be seen as analogue in that they were equivalent to the organic, unfolding and durational processes of humans and their environments and 'whose operations simulated processes that people had seen in nature and in the functioning of their own bodies' (2009: 401). The key point here is that the worlds that humans have constructed through analogue technologies are worlds that they could *recognise* through being modelled on themselves and the environments they lived in. It is a point made famous by Marshal McLuhan in *Understanding Media*, where he argued that mechanical technologies are 'extensions of man' into time, space and nature (1964). From this it's possible to say that modernity itself, in all its diversity, and with its machine and clock-time foundations factored in, was analogue to its core. And it was so most clearly in its specific relationship to the natural world, the world that modern man tried to mimic in his technology development, and that 'industrial man' with his newly acquired modern powers, tried relentlessly to master.

The Analogue and the Digital

Such analogising might seem to be a banal or obscure point. And it would be had it not been for the rise to domination of the antithesis of the analogue— the digital computer, and its suffusion of every register of life. For his part McLuhan was clear that digital technologies, like analogue, were recognisable in nature and part of nature. He wrote, for instance, in his 1970 *Counterblast* that 'The new media are not bridges between man and nature - they are nature' (p.14). For all his insights McLuhan was a philosopher, and very much an epigrammatic one at that. However, regarding the qualities of digital technology in comparison with analogue, we might consider more carefully the opinion of the experts, such as the computer scientists and neurophysiologists who participated in the Macy Conferences on Cybernetics held in New York between 1946 and 1953. At the eighth conference in 1950, Ralph Waldo Gerard, a behavioural scientist, considered the differences in a paper titled 'Some Problems Concerning Digital Notions in the Central Nervous System'. Gerard argued that analogue and digital processes both functioned in the human brain, but that each expressed a different logic. He wrote that:

> [A]n analogical system is one in which one of two variables is
> continuous on the other, while in a digital system the variable
> is discontinuous and quantized. The prototype of the analogue
> is the slide rule, where a number is represented as a distance

> and there is continuity between greater distance and greater number. The prototype [of the digital] is the abacus, where the bead on one half of the wire is not counted at all, while that on the other half is counted as a full unit. [...] In the analogical system there are continuity relations; in the digital, discontinuity relations (1953: 172).

'Discontinuity' is the key point when we speak of technologically metered time and the effects of this upon the individual and society. Clock time is unfolding and continuous and corresponds to nature and to the phenomenology of 'becoming'. Clock time, throughout modernity, has given analogue shape to our understanding of 'cause and effect', which is important in the political process: 'Why do things happen? And what stems from certain causes?' Digital (network) time, by contrast is discontinuous and it is much more difficult for us to discern the 'unfolding' of events, especially at high speed, and likewise to discern cause and effect. Macy experts such as Gerard saw 'discontinuity relations' as little more than an interesting characteristic of digital logic. In the 1950s very few thought that computing would become what it became, and still less could many have anticipated, beyond science fiction writers, that digital computing would dominate society so comprehensively through its capacity to transform technology and network communications.

Since the 1980s this *different category* of techno-logic, one that was subordinate to analogue, has become dominant—with effects we have yet to fully understand. As Arthur C. Clarke (1973: 21) speculated in the early 1970s, the computer is perceived by humans at some level of consciousness as a kind of 'magic', so different is it from any previous technology. It transforms (or consigns to obsolescence) existing technologies, and it creates whole new (virtual) realms of human experience. This is a radical qualitative change that we do not appreciate enough in terms of its pervasive temporal and political effects.

The networked computer's key effects are acceleration and automaticity. The former functions at rates far beyond human cognitive and physical capacities in the communicative process; and the latter is a conscious *engineering out* of the human element from computer processes, a more recent flowering of the cybernetic discoveries that began in the 1940s. Both logics serve to de-link humans from the process of making and communicating—from the 'material forms and processes' through which humans functioned in a world they could recognise. The virtual worlds of acceleration and automation create worlds through means we cannot see, through functions we do not control, and through effects that are increasingly unanticipated.

Such has been the rapidity and extensiveness of computerisation, and such has been our collective faith in their problem-solving capacities and life-enhancing potentials, that not only have we not realised that an analogue world was being eclipsed by a digital logic that infiltrates almost every aspect of economy, culture and society, but we have missed something important about ourselves: *that we are analogue people who are becoming trapped in a digital world*. And it is only now with the domination of digital processes in our lives that we have cause to reflect upon and consider the contrast between two competing technological and ontological states.

What does it mean to be analogue creatures in a digital world? And more importantly how do analogue institutions such as the processes of liberal democracy function in such a context? As we will see, the nature of time can provide answers to what still might seem to many a bizarre couple of questions. Yet the premise to the questions is not bizarre at all if you think about it—which is precisely the problem—we've never had occasion to contrast analogue with digital as an ontological question. Now we do. If we accept the definition of analogue having its basis in nature, and if the technologies we have created have their analogue in nature, then as part of nature we cannot easily exist outside of this interaction; we ourselves *are* nature and therefore analogue beings.

If we care to look, then many important aspects of human endeavour can be seen as analogue processes confronting the challenges of displacement or colonisation by digital logic. For example, the analogue technology of writing, and its mass migration from print to screen from primary school onwards, is a problem in respect to how we understand cognition. What we are and how we think is closely bound up with the technologies we use, and the possible consequences of the radical difference between print and digital writing has hardly been considered beyond marginal voices in remote corners of the academy.

The political process emerges from writing, and it follows that this too is (in its modern forms) irreducibly analogue. The democratic processes to which we still look to provide the means through which individuals and collectives create and enact their political rights and responsibilities within and between nations drew its philosophical energy from the 18th century Enlightenment. And these processes were entimed by the contemporaneous technologies of communication (print and machine and clock) that formed the basis of Debray's graphosphere.

A useful way to think about the modern political process and its temporality is through the usually derogatory term 'machine politics'. Originating in late-19th century USA, the term denoted centralised and hierarchical communicative forms, with long-standing connections of political control, and with outcomes that were planned and often predictable. Usually machine politics is considered an undemocratic corruption: think Japanese politics from the 1950s to the present day. Nonetheless to consider institutionalised political processes as *machine-like* does capture its essential elements—elements that can be democratic, or at the very least have affects that may be seen and recognised and have their own unfolding in time and space—for good or ill, or something in between.

We see the rhythms of the political machine still at work in what are the relatively *slow motion* processes within the institutional spheres, from parliaments and congresses, to council chambers and constituency meetings. We see it in the workings of a modern system of control and of planned outcomes in the analogue rhythms of individuals convening to speak and discuss, of voting cycles, of calendar-driven parliamentary and congressional sessions, of research departments that gather information with which to reflect and project, of committee meetings, of published debates, of scheduled 'hearings' through which face-to-face and often conflictual interaction takes place, and of government bureaucracies that have set administrative procedures through which the political process is filtered and enacted.

These are the 'forms and processes' of Debray's graphosphere in practice, with analogue people expressing ideas through analogue technologies (writing, print, clock and calendar) that functioned as a recognisable process through time and space—and though which people could enact politics to rhythms that (broadly speaking) reflected those of nature.

Traditionally, this machine-like activity had two specific outcomes. The first is *impact*. This is the very point of the political process—to have an effect upon its surroundings and to transform it at the local and/or global scale. The process is of course fraught and imperfect and subject to all kinds of power contestation. Moreover, the impact of politics may be more or less successful, and may be dependent upon how success is measured at the time, and this measure may change with the passage of time.

The second outcome is more interesting for our purposes. This is the creation of a political *environment*. By trying to transform its environment, the political process creates its own, one that is analogous to nature and, more, was intended to improve upon nature (improve our lives) in tangible and recognisable ways. The political environment was the sphere wherein the philosophical Enlightenment idea of *progress* was created and given practical expression in all kinds of ways since the 18th century. The political environment was thus one where an idealised version of the natural world, a Utopia, could be dreamed of, planned for, and sometimes made actual. This political environment unfolded in time and space through what Paddy Scannell (2007:17) called a 'common public time' and this was organised in an analogue mode that people could also recognise because it was a mode that corresponded to the natural world (more or less) and its natural rhythms.

The analogue world created by analogue beings though analogue processes and analogue technologies was so 'natural' that we have barely considered this cornerstone of modernity. However, the digital computer, the machine that David Jay Bolter (1984: 15-40) termed the new 'defining technology' of our post-modern age, has upturned these assumptions.

Preternatural Machines and an Unnatural Political Process

To consider the computer as 'non-analogue' is perforce to consider it *preternatural*, that is to say it's 'beyond nature'. And if it is, then the computer is unprecedented in its importance and is a machine possessed almost of the magical properties to which our species has always been in awe of and susceptible to.

Today we bow before and offer propitiations to a logic we don't really understand and to processes (such as acceleration and automation) that do not exist in nature. The computer's preternatural power stems from the fact that its digital immediacy (as an effect) transcends and makes obsolete the mechanical speed that comprised the speed limit of modernity and the institutional political processes it gave rhythm to.

The networked computer pulls us sharply away from the outmoded modern sphere and its forms and processes. For the digital natives this old world is an increasingly sepia toned one anyway. But it is a world where the analogue institutions and processes of politics continue to function as though its impact and environment-creation capacity is unchanged. Nonetheless, analogue political time has found itself 'out of synch', as Sheldon Wolin (1997: 47) predicted it would be, with the high-speed temporality of the digital network. This has created a profound temporal contradiction. Politics, in its deep-level forms and processes—if not at its surface level of networked communications—continues as if the world were still analogue, as if it were still 1972; a world prior to the computer revolution and where Zhou Enlai's *longue durée* could still shape the political process, and it could do so because analogue was all there was.

To understand the scale of the contradiction between political time and network time, the concepts of mechanical speed and digital speed need to be delineated. The former is analogue, as we have seen, and has speed limits; its processes can be accelerated only so far before they begin to break down. The latter has no inherent temporal limitations, and is constrained only by the current state of technological capacity, which continues to increase exponentially. Social acceleration and democracy don't mesh particularly well. And as digital communication becomes the basis for the articulation of essentially analogue processes, then the ideas expressed in the political process, and of the democratic process in particular, become both *displaced* and *deformed*. It is the negative effects of this asynchronicity that gives form and content to the preternatural political process that rises to dominance in our post-modernity.

Perhaps the most consequential displacement occurs in the realm of political ideas and their mediation. With the ascendancy of neoliberal thought in the late-1970s, political ideas have become increasingly refracted through the logic of capital. And as Marx recognised a long time ago, capitalism thrives upon growing acceleration, and that technological development gets its dynamism from capitalist competition (time is money). The growing dominion of digital acceleration means that those ideas able to synchronise with the ever-faster pace of economy and society are those that find currency most easily.

Jodi Dean sees the displacement of the traditional political realm by digital capitalism as an expression of the power of 'communicative capitalism', where 'the market becomes the site of democratic aspirations' (2005:54). Ideas that look to the longer term, or reach back reflectively to the past for guidance, or ideas that simply need time to articulate and develop properly in order to reflect the inherent need of the idea itself, become marginalised as they literally take too much time. When time becomes oriented towards instrumental 'efficiency' and when the mantra of 'faster is better' is largely unquestioned, then analogue politics becomes increasingly ineffective. It is no longer able to create its own environment wherein it can create and act upon a world that is equivalent to its own analogue capacities.

The machine and clock based 'cognitive framework' that was the diverse and agonistic framework that formed modern political ideas has been captured by the singular ideology of the marketplace logic and the temporal acceleration that is valorised within it. It is in this context of displacement that we can now perhaps understand Perry Anderson's lament that we live in a monopolitical world, where for the first time since the Reformation 'there are no longer any significant oppositions—that is, systematic rival outlooks—within the thought-world of the West' (2000:17).

The politics of modernity—where political forms and processes have recognisable aims and objectives, recognisable as analogues of nature to be worked upon in the creation of a world made better—are being eclipsed. As the analogue political process becomes displaced and marginalised though the negative political power of digital acceleration, *deformations* of the modern political project fan out and proliferate to construct the unnatural politics of our digital post-modernity.

Post-modern Deformities

A post-modern political reality is that social acceleration through digital communication has meant that power percolates up to the level of the corporate boardroom and the political executive. Corporate capital is more powerful than ever before, and CEOs and senior executives need to respond rapidly to economic challenges and opportunities. This much is clear. Less apparent is that in our fast-paced globalised economy and society, the same time-pressures confront political institutions, and their analogue pace is not up to the task. Slow-paced political processes continue in their daily work, of course. Institutions with their committees and legislatures and party structures carry on grinding out platforms and policies and laws. At this level and at this analogue rhythm, politics carries some social meaning still and can function within a recognisable analogue context.

But at the executive level, at the level of the 1% where politics, capital and power intersect, politics essentially flies blind. Political crises and economic imperatives are confronted in hasty and improvised and unrepresentative fashion. The executive is able to act with dispatch—but only to make rapid (and often poorly thought-out) decisions on our behalf. Alexander Hamilton, US founding father, recognised the innate folly in rushing the political process when he wrote that, a 'promptitude of decision is oftener an evil than a benefit' (Hamilton 1788).

An example of this was observed in the opening phase of the 2008 global financial crisis where stock market time pressure for governments to act rapidly was extreme. In September of that year, as the US economy seemed to be spiraling down, the U.S. Senate considered the first of its 'stimulus packages' for its stricken economy. The prevalent feeling was that officials had *only had a weekend* (before the stock exchange reopened) to fix the problem. Senator Lindsay Graham, a Republican representing South Carolina, and present at the crisis meetings declared to a Fox News journalist afterwards that: 'The process that's led to this bill stinks. There is no negotiating going on here! Nobody is negotiating! We're making this up as we go!' (Graham, 2009). When markets are waiting impatiently for signals, and when individuals are not in possession of all the facts nor the necessary time to comprehend them, then 'making it up' is all that can be done.

This is politics conducted in post-modern network time. It occurs in a world of acceleration wherein immediacy governs, not reflection, debate and reasoned decision-making. Power is abstracted and constantly shifts and dissipates and then concentrates and dissipates again, mirroring the random forces of the marketplace where nothing is certain and nothing can be held to proper political and democratic scrutiny. Paul Virilio (1995) termed such a context, a 'dictatorship of speed'.

Perhaps most ominously the alienated and inauthentic politics of network time flourishes among the 99% also. The 'masses' have reached out to information technologies as substitute means of political communication because the analogue political parties and institutions of class based organisation are either dependent upon the neoliberal consensus or are depleted of political power and/or credibility. The 99% no longer have a recognisable 'other' against whom struggles may be conducted, nor political institutions in which they can invest.

The buds of the 2011 Arab Spring, for instance, seemed to portend a political flowering through social media, a Facebook revolution where the digital was adapted for democracy. But its millions of activists were thrust into an accelerated sphere where their Enlightenment-derived aspirations were too far out of synch with both the temporality of the ideas and the political realities of the region. The political analogue of nature's 'grassroots' could not find the soil in which to strike, nor the time for their cultivation. There were in fact no buds because there is no analogue for soil in the network. Here as in all putative 'digital politics' there is, as Moisés Naím (2014) observes:

> a powerful political engine running in the streets of many cities. It turns at high speed and produces a lot of political energy. But the engine is not connected to wheels, and so the 'movement' doesn't move. Achieving that motion requires organisations capable of *old-fashioned* and permanent political work that can leverage street demonstrations into political change and policy reforms. In most cases, that means political parties.

Political parties that aspire to be democratic and connected to the people are indeed 'old-fashioned'. They are also, necessarily, analogue. Old-fashioned parties can be reactionary, too, of course, and those that were organised, as in Egypt, and were disciplined with intact institutions of economic, cultural, political and military power, were able to sweep aside forms of digital organisation that dissipated as rapidly as they formed.

Analogue Epilogue?

Like many artists, Václav Havel saw society at a slightly different angle from the rest of us. Often the artist's insights are perspicacious. Also a democrat and philosopher living in Stalinist Czechoslovakia, Havel was acutely aware of the political role of temporality and of communication technologies. Time in prison, time in waiting and organising, perhaps confident that such a system could not endure over the *longue durée,* was an Enlightenment relationship with time that was analogue. In the pre-1989 Czech graphosphere communication between dissidents was fully analogue also: *samizdat*-based that was conducive to thought, debate, reflection, planning and establishing and nurturing political roots within the idealised environment that Havel and others created.

Havel did not naively imagine a Utopia in the West, however. In the late 1970s, time and distance from the West had enabled him to see where western 'post-totalitarian' societies were headed through emergent information-based technologies. And he could recognise the coming inability, though 'planetary technology' that is 'out of control', for humans to recognise themselves in nature:

> What we call the consumer and industrial (or post-industrial) society…is perhaps merely an aspect of the deep crisis in which humanity, dragged helplessly along by the automatism of global technological civilisation, finds itself (Havel 1978:55).

And post-industrial (or what I've termed post-modern) politics give no protection against the onslaught of digital automatism. Havel continues:

> It would appear that the traditional parliamentary democracies can offer no fundamental opposition to the automatism of technological civilisation and the industrial-consumer society, for they, too, are being dragged helplessly along by it (pp.55-56).

Havel could see clearly the symptoms but not the cause. How could he? The dualism of analogue and digital is only now emerging with the latter's ascendancy. But he could see the threat of technology upon the political process. Looking back it is possible to see that Havel and millions of other thinking, reflective people, West and East, were living in *authentic analogue time* and in an analogue political process in its end phase. We have the contrasting force of digitality to enable us to see that now—and to understand what knowledge of it allows us to see. We are now able to see the value of the analogue relationship with time, and to consider ourselves as analogue creatures that are part of nature. And we can also appreciate, at least vaguely, what the digital universe deprives us of—an authentic (at least in terms of analogue time) relationship with political process so deeply shaped by technology. Even the Stalinism that Havel fought against was temporally authentic, in its way, as was social democracy or old-fashioned conservatism.

Salutary to consider that unless we make the temporal within the political salient—and make efforts to give it back over to human and analogue agency—then the modern political process will slip further towards a state of acceleration and automation, and thence from any potential for democratic control. It would be a terrible and ignominious end for a modern political process (and project) to be eclipsed by technological speed and market imperatives. But, to quote the poet Lawrence Joseph from his 'Visions of Labour', this would be a mistake of our own manufacture: 'Makers, we, of perfectly contemplated machines' (Joseph, 2015:17) who will have become unable through lack of time to contemplate what we have actually made.

References

Anderson, B. (2007) *Imagined Communities.* London: Verso.

Anderson, P. (2000) 'Renewals.' *New Left Review.* No. 1: 5-24.

Bolter, J. D. (1984) *Turing's Man: Western Culture in the Computer Age.* Chapel Hill: The University of North Carolina Press.

Connolly, W.E. (2002) *Neuropolitics.* Minnesota: Minnesota University Press.

Clarke, A.C. (1973) *Profiles of the Future.* New York: Harper & Row.

Dean, J. (2005) 'Communicative Capitalism.' *Cultural Politics* 1(1): 51-74.

Debray, R. (2007) 'Socialism: A Life-Cycle.' *New Left Review* 46 (July–August): 5–17.

Estévez, S. (2009) 'Is Nostalgia Becoming Digital?' *Social Identities* 15(3): 393-410.

Fukuyama, F. (2011) *The Origins of Political Order.* New York: FSG.

Gerard, R.W. (1953) Macy Conferences 1946-1953: The Foundation of Cybernetics. Available online at: http://pcp.vub.ac.be/books/gerard.pdf

Goodman, D. (1994) *The Republic of Letters.* Cornell University Press.

Graham, L. (2009) 'This Bill Stinks … We're Not Being Smart.' *Fox News Interview Archive*. February 6. Available online at: http://www.foxnews.com/story/0,2933,489007,00.

Havel, V. (1978) *The Power of the Powerless.* London: Routledge Revivals.

Havel, V. (1990) *Disturbing the Peace.* London: Vintage.

Hamilton, A. (1788) Federalist Papers # 70. Available online at: http://www.constitution.org/fed/federa70.htm

Hutchings, K. (2008) *Time and World Politics: Thinking the Present.* Manchester: Manchester University Press.

Joseph, L. (2015) 'Visions of Labour.' *London Review of Books.* Number 12

McLuhan, M. (1964) *Understanding Media.* New York: McGraw Hill.

McLuhan, M. (1970) *Counterblast.* London: Rapp + Whiting

Naím, M. (2014) 'Whey Street Protests Don't Work.' *The Atlantic*. April 7. Available online at: http://www.theatlantic.com/international/archive/2014/04/why-street-protests-dont-work/360264/

Scannell, P. (2007) *Media and Communication.* London: Sage.

Scheuerman, W. (2004) *Liberal Democracy and the Social Acceleration of Time.* Baltimore: Johns Hopkins University Press.

Thrift, N. (1996) *Spatial Formations.* London; Thousand Oaks, CA: Sage.

Virilio, P. (1995) 'Speed and information: Cyberspace alarm!' *CTheory*. 27 August. Available online at: http://www.ctheory.net/text_file.asp?pick=72

Virilio, P. (2012) *The Great Accelerator.* Cambridge: Polity.

Wolin, S. (1997) 'What Time is it?' *Theory and Event.* 1(1). Available online at: http://muse.jhu.edu/journals/theory_and_event/v001/1.1wolin.html

7

Time, Power and Inequalities

VALERIE BRYSON
UNIVERSITY OF HUDDERSFIELD, UK

Until recently, political theorists and scientists have had little to say about the political significance or nature of time. In contrast, this chapter argues that the ways in which time is used, valued and experienced are neither inevitable nor 'natural'; rather, they reflect and sustain deep-seated patterns of power and inequality. The analysis of time is, therefore, central to the understanding of political processes and outcomes.

Written from the perspective of a British feminist political theorist, the chapter brings together some recent developments in 'malestream' political theory and international relations, critical discussion of time-use studies and feminist analyses of social reproduction. The starting point is post-industrial societies, seeing inequalities within these as intertwined with intersecting inequalities elsewhere in the world. It focuses in particular on the uneven distribution of 'free' time, the unequal value attached to different patterns of time use and the prioritisation of the 'time is money' culture of both the capitalist workplace and the global economy over other temporal rhythms and needs. The chapter argues that our current relationship with time is both unjust and damaging. It extends feminist insistence on the economic importance of informal, domestic and/or unpaid work to explore the 'time culture' involved in reproductive work; unlike more mainstream writing on the political implications of the accelerated time culture of global finance capital, it is concerned with the uneven impact of temporal changes on unequally situated groups, and the subordination of 'other', more 'natural' time cultures to capitalist imperatives.

'Free' Time as a Scarce and Inequitably Distributed Resource

In his later (2001) work, the eminent political philosopher John Rawls identified leisure time as a 'primary good', something that everyone wants; it is also clearly a scarce resource, and he argued that its distribution should therefore be regulated by principles of justice. This means that those who choose leisure over work should expect to pay an economic price, so that there would be no place for either the idle rich or workshy welfare claimants in a justly organised society. Although Rawls did not make the point, free time is not simply desirable in itself, it is also a key political resource, for any political activity takes time, and those who toil without respite cannot choose to attend meetings or become involved in community campaigns, let alone stand for political office or travel abroad; any inequitable distribution of free time is therefore doubly unjust. Rawls focussed on justice within societies, but in an era of global interdependency it should logically apply also to the distribution of scarce resources *between* societies; clearly, such an extension has profound and radical implications for global justice.

Historically, unequal access to free time has frequently reflected and reinforced unequal access to economic resources, with lower status or class groups and manual workers putting in longer hours for less reward than large land or capital holders and managerial and professional employees. Today, some analysts argue that this class pattern has been reversed in many western nations, with long-hours working and lack of leisure seen as both a sign of status and a means to career success, so that the 'money rich' are now also the 'time poor' (Gershuny 2005; Sullivan and Gershuny 2004). Nevertheless, money can also 'buy' time, for example by taking a taxi rather than waiting for a bus, or by 'outsourcing' family responsibilities. Here, the concept of 'discretionary time', the time remaining after deducting the hours an individual *needs* to spend in paid work, unpaid work and personal care in order to keep out of poverty and meet minimal social standards, may be more useful than free or leisure time, as it helpfully highlights the choice that underlies much long-hours working in the West, whereby individuals work long hours in order to maximise their income and enjoy an affluent lifestyle (Goodin et al., 2008; Burchardt, 2013). Badly paid workers obviously have little scope for such trade-offs, their working hours are generally less flexible and they are more likely to involve anti-social working hours, while the outsourcing of their own domestic and caring work is not an option for those badly paid (usually women, often migrant) workers who provide services for wealthier groups and who may have to leave their own families or neglect their needs (Ehrenreich and Hochschild 2003; Litt and Zimmerman 2003; Peterson 2003, 2012; Runyan and Peterson 2014). Problems are often compounded for the growing number of part-time and/or temporary workers with unpredictable hours and pay: in 2015, an OECD report showed that 'non-standard' work now accounts for around a third of total employment in industrialised countries, and that households depending on these wages are more likely to be living in poverty (OECD 2015); some commentators now refer to those trapped in such insecure, unpredictable employment as the 'precariat'. Other issues arise for those working night shifts, who can be effectively excluded from the social and political life of their community and whose numbers are rapidly rising with the growth in international call centres; here A. Aneesh shows the global power relations involved, as Indian workers have to adjust to the time-zones of clients in the United States, the United Kingdom and Australia and their 'concrete social and personal lives are subordinated to global system imperatives' (2012: 515). Although unemployed people might appear to have more than their fair share of leisure, their ability to take advantage of this is often limited by poverty, and also by the time-consuming nature of welfare application processes and job applications and/or the debilitating psychological effects of unemployment; many unemployed people also have time-consuming caring responsibilities.

Like most male political theorists, Rawls did not explore issues of justice within the home (see Okin 1990; Baehr ed., 2004; Abbey ed. 2013), and he equated leisure time with time left over from paid employment. The view that citizens should contribute to their society primarily through paid work reflects the dominant economic and political assumptions of global capitalism, but it has been challenged in different ways by both liberal and socialist feminist political theorists (for an overview, see Bryson 2016, Chapter 14), and also by feminist scholars in Development and International Political Economy, such as Diane Elson and V. Spike Peterson. As these writers argue, the focus on 'productive' work loses sight of other forms of necessary work, not only the unpaid agricultural work central to subsistence economies, but also the domestic and caring work that is essential to the survival and wellbeing of all societies (most obviously, we all need care when we are born, and also if we become sick or frail). Because this work is often unpaid, it disappears from statistics on national production; however, an international feminist campaign succeeded in committing signatories to the *Platform for Action* that resulted from the 1995 World Conference on Women in Beijing to developing time-use studies to measure unpaid work and include it in national accounts.

The resulting studies have confirmed feminist claims both about the importance of unpaid work and women's disproportionate responsibility for it in all societies (Gershuny 2011), leading many to claim that men's 'domestic absenteeism' underpins the gender gaps in pay, career progression and political representation that remain in all nations (for overviews of feminist arguments, see Lister 2003; Bryson 2007). However, the related claim that women have less access to free time has been disputed by leading time-use researcher Jonathan Gershuny, who argues that in western nations women's longer domestic hours are balanced by men's longer hours of paid employment, so that 'work- and leisure-time totals ... seem generally similar between women and men in each country' (2000: 8-9). Gershuny also identifies some convergence of roles as women have increased their time in the workplace and men contribute more in the home, particularly in relation to childcare. However, Gershuny's finding on leisure time, which uses the evidence of time-use diaries in which respondents are asked to record their activities at fifteen minute intervals throughout the day and night, should be handled with caution. While they appear to provide objective information, these diaries depend on a particular view of time that sees it as unfolding in a series of discrete, measurable activities; this is often alien to the temporal experiences of providing care and it can under-record the sometimes overwhelming impact of domestic responsibilities and the extent to which these affect apparently 'free' time. Thus, some studies record only one activity during any given period, ignoring the fact that, for example, an activity such as 'going swimming' can be transformed from leisure to work if the diary keeper is responsible for young children; many studies have also disregarded the constraining impact of the 'passive' care of a sleeping child or an elderly relative suffering from dementia that does not appear in the diary, but that can prevent their carer from leaving their home. Gershuny's focus on *total* leisure time also ignored the frequently brief, fragmented and unpredictable nature of carers' free time, and the extent to which this leaves men with greater control over their leisure hours and more 'usable' time (for an expanded discussion of these points, see Bryson 2008).

Unequal access to free, discretionary or leisure time is politically important, for it means that those who are most time deprived are least able to gain a political voice. One effect can be the absence of those with direct experience of caring responsibilities from positions of power (and because 'successful' women are disproportionately childfree and without caring responsibilities, a rise in the number of women in decision-making positions does not necessarily mean that the voices of carers are heard). The intersection of time poverty with other forms of social and economic disadvantage means that the poorest and neediest – such as many lone mothers, the 'precariat' and migrant care workers – are particularly unlikely to be heard. Such inequalities are not inevitable, but could in principle be addressed by a range of social policies, including state support for carers, employment policies that assume that 'normal' workers have family responsibilities and welfare policies that recognise that citizens who are not engaged in paid work are not necessarily idle or irresponsible, but may be contributing to society in other ways. As discussed in later sections of this chapter, there have been moves in this direction, but the power of nation-states to plan and implement such changes has been reduced by the global drive to short-term profit and a shift to a new, accelerated relationship with time.

In low-tech societies, the time-consuming impact of household responsibilities is of course often much greater than in the West, and the political exclusion of women much greater (Runyan and Peterson 2014).

However, some writers argue that in recent years the demands of grass-roots women's movements and organisations have been filtered upwards through a wide range of non-governmental organisations to be heard in global forums, feeding into a global shift in thinking about women's human rights, which has had a significant, if patchy, impact at the level of transnational political institutions and conventions (Okin 2000; Walby 2005). In practical terms, however, many apparent gains, such as moves to improve girls' and women's access to education and employment, coincide with capitalism's need for docile, trained employees; such developments leave structured inequalities intact and can look more like 'a depressing example of neo-liberal co-optation' than 'an inspiring example of feminist activism and democratic innovation' (Squires, 2007: 51); as Sylvia Federici argues, claims by the United Nations to promote women's rights 'channel[led] the politics of women's liberation within a frame compatible with the needs and plans of international capital and the developing neoliberal agenda.' (2012: 98. See also Eisenstein 2009)

Valuing Time

While there are many exceptions, high pay and status are generally linked to time spent on non-manual work, with those who sit in offices earning more than those who get their hands dirty, and brain power and education rewarded more than physical strength. This is reflected in a global movement in manufacturing away from wealthier nations, whose workers in service sectors benefit (in the short term) from access to cheap consumer goods made in nations where cheap manual labour is readily available, while financial speculation rather than production is providing the basis for a new, global, super-rich elite. Meanwhile, 'development' and 'progress' are usually understood in terms of a shift away from subsistence activities and towards a market economy, with cash crops and economic growth prioritised over sustainability and all human values reduced to measurable, monetary transactions.

This context compounds the global patriarchal denigration of anything associated with 'the feminine' (Peterson 2012), so that domestic and caring work is often seen as a form of unskilled manual labour or some kind of natural, instinctive extension of womanhood, rather than 'real work'. Those who spend their time on it are therefore economically penalised rather than rewarded; and it often seems invisible to policy makers, who forget that not all value can be expressed in monetary terms, and that apparently 'unproductive' citizens may be making an essential contribution to society. For those who do this work for their own family, it is of course often enjoyable, often immeasurably rewarding, and often undertaken out of love. It is, however, also often physically and mentally demanding, sometimes tedious, frequently exhausting, not always freely chosen and, as discussed above, it takes a great deal of time. Efficient housework and good quality care also require practical and intellectual skills; as with other forms of skilled work, some of these can be learned, and some require natural ability. For example, raising good citizens requires an understanding of child development, and supporting people with dementia requires understanding of their condition; both forms of care also require personal qualities of patience, empathy and the ability to communicate well.

As discussed above, this work is disproportionately done by women, often without pay; when it is done as a form of paid employment it is usually poorly paid, reflecting the low status accorded to both manual and women's work. This low pay makes a major contribution to the global gender gap in pay. Many equal opportunity policies therefore seek to guide young women towards better-paid, traditionally male occupations such as information technology or engineering, and away from traditionally female sectors. At the same time, such policies also now often acknowledge that workplace equality requires childcare provision: for example, a recent report on gender equality by the European Commission argues that 'Europe needs to fully utilise the talents of all its women' (2015: 7) through supporting them into employment, encouraging them away from traditionally female occupation and providing childcare. Quite who is supposed to deliver this childcare if it is deemed too menial for talented and aspiring women is not addressed, but the logical (and presumably unintended) implication is that looking after children in their most formative years should be left to those deemed incapable of 'better' forms of employment or 'outsourced', along with other forms of social care, to female migrant workers, who leave their own families to look after children or frail elderly people in wealthier nations. The more radical option of raising the status and pay of childcare workers (which could have the side-effect of attracting men into the profession) is seldom on the agenda, and the global consequences are ignored.

More positively, gender equality policies today do usually acknowledge that some workers have demands on their time outside of the workplace, and there has been a general movement towards more 'family-friendly' working practices. In particular, most nations now provide some kind of support for employed parents through maternity, paternity or parental leave entitlements (although these are not always realised in practice), leaving the United States among a tiny handful of nations that provide *no* form of paid maternity or family leave for female workers (Rosen 2004). There is also now some recognition of the feminist argument that workplace equality requires that men do more in the home, and by 2013 at least 79 countries provided some kind of leave to fathers around the time of birth (International Labour Organisation 2014). In general, however, the time that citizens need to care for others outside the workplace is still seen as something to be fitted around employment, rather than the other way around, and more flexible, family-friendly forms of employment are still largely viewed as special treatment for women workers rather than the norm. As discussed in the next section, this prioritisation of the values and needs of the workplace also involves prioritising a particular perspective on the nature of time itself, linked in turn to a weakening of the power of nation states and a crisis both in care and in capitalism itself.

Time Cultures

As many writers have now shown, the ways in which we experience and understand time are not straightforward and 'natural' but socially and culturally produced (from a wide literature, see in particular Thompson 1999 [1967]; Adam 1995). Some historians have identified a shift in western societies from a pre-capitalist, pre-industrial time culture that was oriented to the cyclical rhythms of the seasons and tasks that had to be done to the time culture of the capitalist workplace, with its sense of time as something that steadily moves in one direction and that can be owned, measured, saved, spent or wasted. Such commodified clock time can be assigned an abstract monetary value and subject to considerations of cost efficiency; its results-oriented, 'time is money' logic is the time of global capitalism, its effects now exaggerated by digital time, with its attendant notions of multi-tasking, the blurring of the distinction between work and leisure and the acceleration and compression of time, most obviously in the case of financial markets, where fortunes can be made or lost in a nanosecond.

Some writers have built on William Scheurman's (2004) claim that today's 'social acceleration' of time is having detrimental political effects, disrupting the slower temporal rhythms of liberal democracy and strengthens executive power, with its ability to respond rapidly to immediate demands and events, at the expense of the long-term planning or reflection on the past that respectively characterise legislative and judicial power. In particular, Robert Hassan argues that clock time has been superceded by a shift to digital, networked time and that this is linked to a shift in power away from nation states to abstract market and technological forces 'not under any meaningful democratic control anywhere in the world' (Hassan 2012:294). Wayne Hope further argues that the inescapable imperative to pursue high-speed, short-term financial profit is in conflict with longer-term strategies of capital accumulation, producing a crisis of capitalism that is effectively a 'crisis of temporalities' (2011: 194), of which the 2007-8 financial collapse was an early manifestation.

These points are important. However, what Hassan and Hope fail to see is that, while clock time ticked more slowly than digital time, an older, pre-capitalist time that cannot be rationally planned and controlled, inevitably still persists in the rhythms of the seasons and our bodies, and in the often unpredictable patterns and needs of emotional and caring relationships. Today, these older times are increasingly subject to the logic of measurable, commodified time, both clock and digital, as the relentless pursuit of short-term profit extends more vigorously into all areas of life in a world in which 'Infants, human organs, sexualised bodies, intimate caring, sensual pleasures, and spiritual salvation are all for sale' (Peterson 2003: 78). In the case of the natural world, the effects are potentially disastrous in terms of sustainable agriculture and climate change; the effects on our human relationships are also deeply damaging.

Caring for other people is an inherently relational and often open-ended activity, while the tasks it involves are often determined by need rather than by the clock (a child's nappy has to be changed when it is dirty, not because it is four o'clock). Many caring tasks are very much focussed on the here and now, and attempting to speed them up can often be counter-productive (trying to rush a child through eating breakfast may provoke a temper tantrum that makes them even later for nursery). Care is also often repetitive rather than with an identifiable end product (that nappy will soon need changing again); it can involve a jumble of simultaneous activities, emotions and processes (the child may not just be 'eating breakfast', but exploring tastes and textures and developing their speech while learning the effects of particular forms of behaviour); and it can take the form of simply 'being there' in case of risk or need. As discussed above, the impact of such responsibilities cannot be fully captured by time-use diaries, which are inherently tied to measurable clock time.

All this means that the 'temporal logic' that good care requires is fluid, relational, often cyclical, and oriented towards contextualised and often unpredictable needs; this is very different from the results-oriented logic of the workplace that seeks to maximise output while using as little time as possible. However, when unpaid family time has to be fitted in around the demands of employment, it too has to be organised and planned to an extent that can feel like the 'McDonaldisation of love' (Boyd 2002: 466, quoting Anne Manne). Meanwhile, in a neoliberal age of austerity, paid care is increasingly seen as a profit-making industry that seeks to 'process' as many clients as possible and that loses sight of less tangible processes; when this care is publicly provided it is increasingly subject to considerations of cost-effectiveness that focus on measureable outputs and value for money, but that may be counter-productive. For example, care workers supporting elderly people in their home may appear to increase their 'output' if they rush through the process of getting them up and dressed in the morning; but if they are able to make this a leisurely process, involving a chat and a shared cup of tea, they may be enabled to remain in their own home and out of residential care for longer, saving public money in the long run. As with family life, the problem is not simply that people do not have enough hours and minutes to do their caring work, although this is important, but that their activities are being forced into an inappropriate temporal straightjacket based on the logic of market capitalism – that is, an economic system based on the pursuit of profit rather than the satisfaction of human need in which reproductive labour is 'increasingly organised for *and* undermined by neoliberal globalisation' (Runyan and Peterson 2014: 200).

Because women throughout the world do more caring work than men, it is generally they who are most affected by the need to 'straddle multiple temporalities' (Everingham 2002: 338), and the extent to which the demands of the workplace are prioritised over other needs is bound up with the more general privileging of both typically male experiences and life patterns and gendered nature of social, economic and political inequalities. These gendered inequalities intersect with those of class and race within and between nations as the 'care deficit' experienced by some households, communities or nations is displaced onto others, producing a crisis in care and social reproduction that is linked to the more general temporal crisis in global capitalism identified above. In this context, as political theorist Nancy Fraser has argued, it is politically important to re-assert the values associated with reproduction, which are 'pregnant with critical-political possibility' and can provide powerful resources for anti-capitalist struggles (2014: 69. For related arguments, see Peterson 2012; Bryson 2011 and 2016).

Conclusions

Our relationship with time in contemporary societies is neither 'natural' nor just. On the contrary, time is used, valued and understood in ways that reflect and sustain economic, social and political inequalities. These inequalities operate at global, national, local and domestic levels to privilege some nations, classes and ethnic groups over others. In prioritising the temporal experiences and needs traditionally associated with men rather than women and generally providing men with more free time than women, today's time culture also sustains deep-seated gender inequalities in all areas of life. These inequalities and injustices are deeply damaging to the health and welfare of citizens and their families throughout the world. It is in the interests of us all to rethink our human relationship with time, and to confront the conflicting, hierarchically ordered, demands of caring, clock and digital times.

References

Abbey, R. (ed.) (2013) *Feminist Interpretations of John Rawls.* Pennsylvania: The Pennsylvania University Press.

Adam, B. (1995) *Timewatch. The Social Analysis of Time.* Cambridge: Polity Press.

Aneesh, A. (2012) 'Negotiating Globalisation: Men and Women of India's Call Centers.' *Journal of Social Issues* 68(3).

Baehr, A. (ed.) (2004) *Varieties of Feminism Liberalism.* Oxford: Rowman and Littlefield Publishers, Inc.

Boyd, E. (2002) '"Being There." Mothers Who Stay at Home, Gender and Time.' *Women's Studies International Forum* 25(4).

Bryson, V. (2007) *Gender and the Politics of Time.* Bristol: The Policy Press.

Bryson, V. (2008) 'Time-use studies: a potentially feminist tool?' *International Feminist Journal of Politics* 10(2).

Bryson, V. (2011) 'Sexuality: The Contradictions of Love and Work.' In: Jonasdottir A., Bryson, V. and Jones, K. (eds) *Sexuality, Gender and Power. Intersectional and Transnational Perspectives.* New York and Abingdon: Routledge.

Bryson, V. (2016) *Feminist Political Theory.* Basingstoke: Palgrave Macmillan.

Burchardt, T. (2013) 'Time, Income and Freedom.' In: Coote, A. and Franklin J. (eds) *Time on Our Side: Why We All Need a Shorter Working Week.* London: nef [the new economics foundation]).

Ehrenreich, B. and Hochschild, A. (eds) (2003) *Global Woman: Nannies, Maids and Sex Workers in the New Economy.* London: Granta Books.

Eisenstein, H. (2009) *Feminism Seduced. How Global Elites Use Women's Labor and Ideas to Exploit the World.* Boulder, Colorado: Paradigm Publishers.

European Commission (2015) *Report on Equality Between Women and Men 2014.* Available online at: http://ec.europa.eu/justice/gender-equality/files/annual_reports/150304_annual_report_2014_web_en.pdf

Everingham, C. (2002) 'Engendering Time: Gender Equity and Discourses of Workplace Flexibility.' *Time and Society* 11(2/3).

Federici, S. (2012) *Revolution at Point Zero. Housework, Reproduction and Feminist Struggle.* Oakland: PM Press.

Fraser, N. (2014) 'Behind Marx's Hidden Abode. For an Expanded Conception of Capitalism.' *New Left Review* vol. 86.

Gershuny, J. (2000) *Changing Times: Work and Leisure in Postindustrial Society.* Oxford: Oxford University Press.

Gershuny, J. (2005) 'Busyness as the Badge of Honour for the New *Superordinate* Working Class.' *Social Research* 72(2).

Gershuny, J. (2011) *Time-Use Surveys and the Measurement of National Well-Being.* ONS and University of Oxford Centre for Time Use Research. Available online at: http://www.ons.gov.uk/ons/rel/environmental/time-use-surveys-and-the-measurement-of-national-well-being/article-by-jonathan-gershuny/index.html

Goodin, R., Rice, J., Parpo, A. and Eriksson, L. (2008) *Discretionary Time: a New Measure of Freedom.* Cambridge: Cambridge University Press.

Hassan, R. (2012) 'Time, Neoloberal Power and the Advent of Marx's "Common Ruin" Thesis.' *Alternatives: Global, Local, Political* 37(4).

Hope, W. (2011) 'Crisis of Temporalities: Global Capitalism After the 2007-08 Financial Collapse.' *Time and Society* 20(1).

International Labour Organisation (2014) *Maternity and Paternity at Work. Law and practice Across the World.* Geneva: International Labour Office. Available online at: http://www.ilo.org/wcmsp5/groups/public/---dgreports/---dcomm/---publ/documents/publication/wcms_242615.pdf

Lister, R. (2003) *Citizenship: Feminist Perspectives.* Basingstoke: Macmillan.

Litt, J. and Zimmerman, M. (2003) 'Global Perspectives on Gender and Carework: An Introduction.' *Gender and Society* 17(2).

OECD (2015) *In It Together: Why Less Inequality Benefits All.*.Paris: OECD Publishing. Available online at: http://www.oecd-ilibrary.org/employment/in-it-together-why-less-inequality-benefits-all_9789264235120-en

Okin, S. (1990) *Justice, Gender and the Family.* New York: Basic Books.

Okin, S. (2000) 'Feminism, Women's Human Rights, and Cultural Differences.' In: Narayan, U. and Harding, S. (eds) *Decentering the Center. Philosophy for a Multic-cultural, Postcolonial, and Feminist World.* Bloomington and Indianapolis: Indiana University Press.

Peterson, V. S. (2012) 'Inequalities, Informalisation and Feminist Quandaries.' *International Feminist Journal of Politics* 14(1).

Peterson, V. S. (2003) *A Critical Rewriting of Global Political Economy. Integrating Reproductive, Productive and Virtual Economies.* London and New York: Routledge.

Rawls, J. (2001) *Justice as Fairness.* Cambridge, MA: Harvard University Press.

Rosen, R. (2014) 'A Map of Maternity Leave Policies Around the World.' *The Atlantic.* June 20. Available online at: http://www.theatlantic.com/business/archive/2014/06/good-job-america-a-map-of-maternity-leave-policies-around-the-world/373117/

Runyan, A. and Peterson, V.S. (2014) *Global Gender Issues in the New Millenium.* Fourth edition. Boulder, Colorado: Westview Press.

Scheurman, W. (2004) *Liberal Democracy and the Social Acceleration of Time.* Baltimore, MD: John Hopkins University Press.

Squires, J. (2007) *The New Politics of Gender Equality.* Basingstoke: Palgrave Macmillan.

Sullivan, O. and Gershuny, J. (2004) 'Inconspicuous consumption: work-rich, time-poor in the liberal market economy.' *Journal of Consumer Culture* 4(1).

Thompson, E. P. (1999 [1967]) 'Time, Work-Discipline and Industrial Capital.' In: Thompson, E.P. (ed.) *Customs in Common.* London: Penguin Books.

Walby, S. (2005) 'Introduction: Comparative Gender Mainstreaming in a Global Era.' *International Feminist Journal of Politics* 7(4).

8

War through a Temporal Lens: Foregrounding Temporality in International Relations' Conceptions of War

CHRISTOPHER MCINTOSH
BARD COLLEGE, USA

For those focused on questions of security, war poses a paradox. On one hand, the concept of war seems obvious and intuitive. It operates as the organising concept around which contemporary security discourse positions itself. Whether one views the category of security in broad or narrow terms, war, armed conflict, and political violence remain solidly within the frame (Lipschutz 1995). Simultaneously, however, there always seems to be an increasing awareness of how the nature of war itself is changing and shifting.[34] Great power wars and state-state conflicts over control of territory as traditionally understood no longer dominate the international space. Drone strikes, irregular warfare, insurgencies, terrorism, as well as massacres, genocide, peacekeeping and humanitarian interventions occupy increasingly central positions in contemporary security studies (see e.g., Boon and Lovelace 2012, Gleditsch 2013). The idea of war seems to be shifting beneath our feet and the concept itself seems difficult if not impossible to pin down in terms of meaning. The paradox this poses is that despite the ever-present gap between the lived, present-day experience of political violence and its dominant representation as 'war', this concept *still* retains meaning in

---

[34]   I recognise from the outset that this is primarily a matter of appearances and explicitly make the argument in later sections that this disjuncture between actual political practice and conceptual representation is an inescapable one, not, as some argue, a product of particular changes in contemporary warfighting.

theory and practice. Much like terrorism, the concept of war remains nearly impossible to define with any certainty or consensus, yet it still remains resonant and foundational in the field of IR (Cronin 2009). The questions this raises are not merely academic. In Balibar's assessment of the war on terrorism he states:

> It seems to me that many, if not all, of the current discussions are in fact obscured and affected by the...preliminary question which remains undecided throughout...when it comes to either drawing lessons from the war, or deriving consequences, or proposing alternatives, is embarrassing for any speaker, namely, the simple question: *what is a "war"?* I shall try to show that this is not a verbal puzzle, a pure nominalistic requisite, that it has consequences for our reasoning concerning the current situation, and the kind of historical determinations that it reveals (Balibar 2008, 366).

The question of whether or not the United States is at war with the Islamic State, for instance, has significant impact on US behaviour and foreign policy. A debate has recently emerged within the United States regarding the necessity of a congressional resolution authorising the use of military force (AUMF) for the ongoing US military operations in Iraq and Syria. An AUMF is the modern American instantiation of an official declaration of war, and as such, it triggers a comprehensive legal shift throughout US policy. The Obama administration claims that they do not need authorisation on the basis of the 2001 AUMF authorising the war on terrorism and more recently narrowed by the Administration to the war on 'Al Qaeda and its affiliates' (US National Security Strategy 2010). At this point, however, the debate has largely been left behind as it has become clear that no new AUMF can be agreed upon (Leatherby 2015). This has left the administration position in place by default—military operations can continue against the Islamic State, as well as against Al Qaeda and whoever else the Administration decides constitutes 'its affiliate'.

The seemingly simple question—whether the United States is engaged in a war with the Islamic State—is a question that cannot be answered with finality at the moment and is now seen as a legal issue, but the conceptual implications involved in answering the question are much broader, just as Balibar argued was the case with the initiation of the war on terrorism. Answering this question with any precision goes well beyond the narrow issue of legal declarations of war. Two puzzles arise at the conceptual level that this chapter will explore. First, there is the question of how past, present, and future operate in the practice of war itself—e.g., temporal horizons that appear obvious in retrospect are very much not so in the moment. Second, at the analytical level, the beginning and endings of war operate in a much more indeterminate manner than typically understood, especially in the study of IR.

International relations (IR) has a conception of war that tracks largely to societal discourses where war is viewed as involving the use of violence between organised groups for ostensibly political purposes (Sambanis 2004; Harff 2003). What's equally important to the behaviour the concept represents is the level of casualty and magnitude of destruction. We may disagree on the level at which a conflict becomes more than just a skirmish or battle, but we can all agree, for instance, that a war is different than an assassination. Regardless of where we draw that line, assessing whether a given conflict has crossed it cannot be known in advance. In other words, as a category of analysis, a war only becomes a war after the fact. Balibar captures this directly when he states:

> Wars are named: they have to be. Most of the time they are named after the event, by historians, which means in particular that they are named when they are considered to be finished, to have been brought to an end... They are named individually or collectively, e.g., the Wars of the Roses, or the Punic Wars, the Napoleonic Wars. But often they are not named univocally (Balibar 2008; 367).

A shooting at Fort Hood inspired by a terrorist group's propaganda or the state execution of an individual identified as conspiring with that group are indeed acts of violence between groups in conflict, but *alone* neither are sufficient to constitute a war. That designation or 'naming', to use Balibar's term, only arises with certainty in the future. War then seems to operate in a continuing space of temporal liminality where it is both outside and beyond our apprehension in the present, yet simultaneously meaningful in context and seemingly appreciable, even obvious.

IR has seen recent movement towards recognising the importance of temporality in all facets of international political practice by foregrounding it as a stand-alone issue, rather than as an issue of history or methodology (Hom 2010, Hom and Steele 2010, McIntosh 2015, Solomon 2013, Hutchings 2008, Hutchings 2007, Berenskoetter 2011, Stevens 2016). This chapter seeks to build upon these moves by illustrating some of the ways in which attention to the temporal dimension in IR can better inform the understanding of a foundational concept such as war. As with any move to incorporate more 'social' aspects of political experience, moves to foreground temporality run into the criticism that these issues wash out when engaging 'harder' issues such as war and the use of force (Mearsheimer 1995, Wendt 1995). This chapter uses a temporal frame to sketch out some of the important aspects of the concept of war as it is produced and reproduced in IR scholarship and practice. In doing so it also operates as an illustration of what can be illuminated by utilising it as a frame (Butler 2009). Wars are relational—both in terms of space and time—and rely on connections in order to come into existence (Jackson and Nexon 1999). As a concept that is already understood as an event—rather than a thing—war is necessarily produced through processes and relations over time and only comes to be understood through these representation(s). Events are difficult to pin down ontologically, yet most accept that they are distinct from things. MacKenzie identifies this distinction, drawing on Davidson, by emphasising that events are distinct due to the action that is intrinsic to the concept—e.g., 'passing legislation' is an event that may be captured by the word 'legislation' but is not synonymous with the abstract and fixed idea of a 'law' (MacKenzie 2008).[35]

This chapter will proceed in two parts. First, it sketches out some important aspects of war that are more clearly exposed when confronted from a temporal perspective. Framing war as constituted first and foremost temporally, rather than as an entity or outcome to be explained and/or understood emphasises the importance of process, discontinuity, transformation, and duration (Buzan and Lawson 2012, Musgrave and Nexon 2013). The second section identifies some of the implications for future security studies scholarship if war is 'fully temporalized' and understood as a dynamic process. Reflexive criticism becomes increasingly important, as the concept of war is susceptible to reification by the very theorists who are seeking to better understand it. To accept the portrayed representations by political actors as real is to leave a critical element of political violence uninterrogated (Butler 2004, Butler 2012). As well, the concept of war should be more fully critiqued and provisionally understood as a category that captures multiple, disparate types of political violence. As such, it may—or may not—remain useful as time goes on.

Identifying War's Constitutive Temporal Dimension

In IR, war is typically treated as an outcome to be explained or predicted. Do arms races precede conflicts? Are great power wars more or less likely during times of transition? Is expansionism still a meaningful motivation for conflict initiation? In political practice, actors that consider themselves in an area of peace see it as something to avoid, an undesirable outcome, or evidence that the systems in places to prevent it have either failed or been insufficient

---

[35] It should be noted that MacKenzie's concern is with the more specific 'political event' and does not necessarily share Davidson's ontological account of this distinction.

(Bialasiewicz and Campbell et al 2007). Regardless of how one chooses to understand or characterise it, however, war is an event. Like any event, and as Balibar argues, it is understood to have a beginning and an end. This section will address this constitutive dimension in three parts—the past, present, and future. While analytically I make this separation to show how the two puzzles operate in a more appreciable manner, it is *not* the case that there is a neat separation. Past, present, and future are each constituted by the other due to the interpenetration of past and present, present and future, as well as past and future—the separation I use here is purely heuristic.

War's (of the) Past

When considering events in social and/or political terms, temporality and temporal dynamics are especially important—historical sociologists like William Sewell, for instance argue that viewing historical events (like wars) in context requires a recognition that temporal relationships are much more complicated, non-linear, and dynamic than we may otherwise appreciate (Sewell 1996, Abbott 2001). Three aspects are particularly important. First, events and their causes are constituted by indeterminacy. To declare a war was 'caused' by single or even multiple factors requires an elimination of the contingency inherent to any historical event. Each historical event is made up of thousands of actors interacting many times over—the complexity involved is dramatic and deep as Becker observed many years ago in articulating his notion of a 'historical fact' (Becker 1955). Second, events are represented in the future—the Great War only becomes World War I after the fact and only even becomes 'Great' after enough time has concluded for that adjective to apply. Lundborg's work on Deleuze and the 'pure event' in the context of '9/11' demonstrates that events are constituted by the constant interplay between past and future—overlapping, unfolding, and constantly interpreted and re-interpreted in the ongoing present (Lundborg 2012, Hutchings 2008). Third, the temporal dynamics of events reinforce the susceptibility of broad structures to transformation—these transformations can be slow and linear, but equally so they can be quick and abrupt. Given that international institutions are constituted relationally by interactions that produce and reproduce these institutions, international political practice would seem especially susceptible to these. In the context of war, given Clausewitz' observation regarding the inevitability of 'friction', this only seems to be an even more important factor (Clausewitz 1984).

Wars, then, are discursively produced as events with a beginning and an end, possessing temporal boundaries that are articulated as discrete and knowable. Mary Dudziak writes, 'war...has temporal boundaries on both sides' (Dudziak 2012). Despite this, the indeterminacy and contingency

Sewell and others speak to manifests particularly when one seeks to precisely articulate this event's boundaries. In the context of war, this occurs when looking to identify a war's originary moment. Take the example of the war on terrorism (Jarvis 2008). Most would place the beginning date of the war as 14 September 2001—the date Congress passed the AUMF. Some might argue that the Bush address the night of the attacks claiming that the United States would 'win the war against terrorism' constituted its beginning. Regardless of the precise moment, almost no IR scholar or political official would argue it began at any point *before* 11 September 2001 (Bush 2001). What is odd about this is that the adversary had already declared war on the United States many years before and the United States had even responded to these past threats with military operations. Shultz and Vogt argue that this exposed conceptual issues, not merely bureaucratic ones.

> Beyond the refusal of the US intelligence community, and for that matter the military establishment, to classify terrorism as warfare because it was not a serious enough danger, other reasons also contributed to this reluctance. Most important, terrorism was not war because it did not resemble modern war as the spooks and soldiers had known it, studied it and practiced it. Therefore, ipso facto, it could not be war (Shultz and Vogt 2003).

What appears a seemingly simple question—when did the war on terrorism begin—is one fraught by perspective and temporal framing. Recognising Al Qaeda's declaration of war on the United States as the start of the conflict confers upon them a legitimacy that many would be uncomfortable with, but from the perspective of the IR scholar, that discomfort should be irrelevant. To act otherwise is to privilege one actor's viewpoint over the other in their analysis of what appears to be an objective fact—the beginning of the war on terrorism.

Some of this depends upon perspective. From the perspective of the IR scholar, the concept of war is important as a way of categorising an event in order to explain and/or understand it. In other words, IR scholars can wait until the war has concluded and is unquestionably a war due to the duration, magnitude, actors involved, etc. These are all things best assessed in the future with the benefit of hindsight. In practice, however, it is vastly different, as wars are declared (and fought) in a forward-looking manner. But even for the practitioner, the interpenetration of the present and future are at play. A declaration of war in the present, for instance, also operates as an articulation of a particular future. It is a future where there is a claimed willingness to engage in actions that would constitute a war, even if they have not as of yet actually happened. Without that future informing the declaration in the present, the present declaration will not be understood as an actual war. Congress authorising war, for instance, articulates a set of futures that would constitute a war should they come to pass—a commitment to ongoing battles, acceptable levels of casualties, massive resource commitments, and a willingness to use force against the enemy. My declaration of 'war' on the spider occupying an unreachable corner of my office, does not, and is therefore appropriately not considered the beginning of an actual conflict.

The United States' current actions against the Islamic State in Syria and Iraq express a similar indeterminacy. If we consider the current actions to be a war—even if ultimately it remains solely an aerial campaign[36]—when exactly would we describe the campaign as beginning? Equally complicated, what will we be saying it even *is* years from now? And if these 'names' change, then how does one act differently and how do our understandings and theories shift? For instance, future historians could read the past decade and a half of US military operations in Afghanistan, Iraq, and now Syria (as well as Yemen) as part of one broader war rather than separate wars. One could even imagine it being understood as a third world war or the beginning of a new cold war as some—admittedly ideologically motivated—have already done.[37] Equally so, historians might take the view that each of these 'contingency operations' are individual conflicts connected in some ways, but not constituting one continuous event. Legally, the picture remains murky as the US is currently justifying its operations in Iraq and Syria as part of a broader fight against Al Qaeda's affiliates even as it asks for congressional authorization specific to this campaign. All this has occurred even while acknowledging that the Islamic State has publicly split with Al Qaeda and is no longer an affiliate.

War's Present

How the future reads our current present could have significant implications for conceptions of IR going forward and how future scholars and observers interpret the actions of those in the present. Future events have the potential to radically re-shape the understanding of present political practices. If the nuclear deal with Iran fails—as many in the US Congress are hoping—and results in an eventual military campaign against an Iranian nuclear programme—which a startling number of politicians are also hoping for—

---

[36]   Officially speaking this is the case, but historically the United States has employed covert operatives to assist when engaged in these types of measures.

[37]   US Secretary of State John Kerry somewhat unwittingly encouraged this in comments comparing the ease of victory in the Cold War to the conflict with the Islamic State. (Sydney Morning Herald 2015)

present day events could look very different to scholars of the future, even though nothing has actually changed. Dick Cheney, for one, is already linking the 2003 invasion of Iraq to the effort to eliminate Iran's nuclear programme, a position that could seem more reasonable if the US were to eventually begin a military conflict with Iran (Carter 2015). Under that scenario, it would become much easier to conceptualise the broad sweep of these events as part of one larger narrative. Previous operations throughout the region could be conceptualised as the beginning of one broader fight against Iranian-backed fighters or as the emergence of a Shia-Sunni conflict. Some parts of the American political scene already seem to believe this is happening (see e.g., Jamie Dettmer 2015). But again, much of this depends almost entirely upon where one locates the origin of the war—regardless of which war we are speaking about.

Separately, if these conflicts are seen as part of one broader campaign (a position the US is admittedly unclear upon) and they are also accepted by future scholars as prompted by the 2001 terrorist attacks, one could easily imagine literature comparing the causes of world wars I and II or Vietnam to the causes of this war given the time the war has taken as well as its (potential) scope. One can imagine under that scenario that the attack on Archduke Ferdinand becoming more broadly understood as a terrorist attack, prompting comparisons to the attacks on the World Trade Center and the Pentagon in 2001. Audrey Cronin, for instance, has already asserted this comparison (Cronin 2009). Alternatively, if it is portrayed as a Sunni-Shia conflict as some argue, it could be compared to the Cold War where two entities are in conflict but refuse direct engagement except via 'proxy wars' (e.g., Yemen) or covert operations supporting rebels (Syria). All that said, it's also possible that each conflict could continue to be understood as completely separate where the US invasion and occupation of Iraq is a single conflict, as was the war in Afghanistan, and as are operations in Syria/Northern Iraq, each possessing their own causes and effects.

How one resolves these issues depends on a variety of concerns including ideology, data interpretation, historiography, and viewpoint and my intention is not to preemptively resolve the inevitable historical debate. But viewing the temporal aspects of this issue exposes two issues. First and foremost, present-based analysis of the conflict occupies a liminal space where the future and past are interpenetrated but contingent. If we believe the conflict is going to end soon our understanding of what it is that we are explaining and/or understanding perhaps goes in one direction, but if we assume the conflict's endpoint is in the distant future, it may be quite another. This leads me to the second point: this is not merely of concern for the scholars seeking to refine the analytic category of war in post-conflict theorisation, but it also shapes the actions of practitioners in the moment. This distinction is an important one, but not limited to this particular context, because the temporal experience of the observed almost necessarily differs from that of the observer. Bourdieu identifies this as a 'de-temporalising' move where

> Scientific practice is so 'detemporalized' that it tends to exclude even the idea of what it excludes: because science is possible only in a relation to time which opposed to that of practice, it tends to ignore time and, in doing so, to reify practices...The detemporalizing effect (visible in the synoptic apprehension that diagrams make possible) that science produces when it forgets the transformation it imposes on practices inscribed in the current of time, i.e. detotalized, simply by totalizing them, is never more pernicious than when exerted on practices defined by the fact that their temporal structure, direction, and rhythm are *constitutive* of their meaning (Bourdieu 1977).

The observation Bourdieu makes here is an important one, precisely because the shift in level of analysis from practitioner to scholar is one that intrinsically creates this 'detemporalising' effect. One can see this when approaching the near-past and present of US foreign policy regarding the 'war on terrorism' as broadly understood. Actions like the surge in Iraq or Afghanistan are articulated as a means of building up indigenous capacity so as to better enable a drawdown of US forces, but when read against the backdrop of a decade and a half with more engagement than not, they can appear as something else entirely—escalation rather than a de-escalatory move. Similarly, ideas like the Obama administration's claim that the United States has 'pulled out of Iraq' at one point seemed quite viable, but given the current conflict and military operations in Northern Iraq and Syria the claim becomes much more complicated. From an IR perspective this complicates even basic questions of cause and effect because of the difficulty in assessing the temporal boundaries regarding when a conflict has started and when it ended (Dudziak 2012). In some ways this might point to a reason why objectivist IR largely ignores ongoing conflict and chooses to deal with the more distant past.

What we think of as the war itself—the duration or middle of the event where the actual armed conflict takes place—is also an intrinsically temporally constituted activity that requires projections into the past and/or future. A war only becomes indisputably a war after the violence reaches a level of significance commensurate with extant discursive understandings of a war. A preemptive strike on a terrorist cell by US operatives in Somalia in and of itself does not constitute a war even though it may constitute an 'act of war'.

Wars are not constituted by a single act, however, but by a series of such acts understood in relation to each other. While this begs the question of how many 'acts of war' it takes to become a war—echoing the morally bankrupt distinction the Clinton administration famously made in Rwanda between 'acts of genocide' and 'genocide'—the lack of an objective answer does not mean that there are no answers accepted and defended in practice (Power 2001). Put differently, the future has a potentially defining role in conceptualising the actions of the present and past. Firing a gun across a border in and of itself does not constitute a war—even if it starts one—unless events that transpire afterwards can be read in conjunction with that act to make it so. The Great War only becomes World War I as a result of future events, but the argument here runs deeper than that. The first shot (as well as the second, third, and so forth) must be followed by other acts in order to be understood as the 'first' shot in a war and that is something of which one can only be certain of (if ever) in the future.

The present of war, then, is temporally complicated and dense—the fog of war extends to the understanding of time and how it transpires for those in the midst of conflict. World wars do not begin as such and participants are so busy in the trenches (so to speak) that taking time out to step back and assess the situation as the historians will is unlikely and perhaps even counterproductive. The temporal present for these actors is constituted by the interplay of incomplete representations of past and future. Applied to today's events it means in the future, when seeking to explain or understand the actions of practitioners of the present, we should take on the temporally appropriate perspective and resist the determinism that comes along with the privileged viewpoint of looking backward.

War's Future

Making sense of temporal boundaries is difficult at the outset of war, but similarly complicated when identifying the end point. When evaluating how wars end, three elements of temporal dynamics complicate thinking on the causes of conflict and cooperation. First, war appears largely as a representation of the past or projection of a particular future, and is not an objectively constituted, independently verifiable entity. While terrorism has seen a great deal of investigation into the term's discursive construction and the power relations at play when identifying an act as 'terrorist' or a group as engaged in 'terrorism', there has been less of this with respect to the concept of war itself (See e.g., Cronin 2003, Hoffman 2006 and Jackson 2007).

Generally speaking, IR seems to rely on a notion of 'we know it when we see it' when it comes to identifying wars. As a result, IR's view tends to reflect the viewpoint of practitioners, particularly amongst objectivist scholars. Naming a set of actions of political violence as war—rather than a massacre, civil conflict, terrorism, or police action, for instance—reflects dominant power relations and the privileged place of sovereign states as international subjects. For the scholar of IR, analytically identifying a set of actions as a war occurs in the present and extends backwards, much as Wendt and others have argued (Wendt 1999, Berenskoetter 2011).

Second, and primarily because the notion of war itself relies upon understandings of the past and/or future, it is best understood as constituted by the intersection of both. David Weberman and Huw Price have both articulated the possibility of 'backward causation' in sociopolitical life. Weberman does this philosophically and Price from the perspective of quantum physics (Weberman 1997, Price 1996). Each ultimately identifies how events of the future can actually change things that have happened prior—physically Price sees this as theoretically possible at the quantum level and Weberman identifies how actions taken in the moment have effects later on that re-constitute the initial action as something different, such as an assassin's bullet only becoming part of the assassination itself after the victim dies. This outcome is not always co-terminous with the shot itself. Both of these analyses echo Lundborg's (2012) analysis of '9/11' to show how events become increasingly interrelated temporally the finer one's focus goes. Independent of how one resolves these issues or even whether one accepts the potential of 'backwards causation', an inevitable aspect of conceptualising a set of events as a war relies on the future to shape the understanding of the past that operates in the present. In other words, to return to the war on terrorism example, we could be in the midst of what will be known as something much bigger than 'contingency operations' (perhaps the first half of the second Thirty Years' War), but do not know it as of yet, because future events have not taken place that make it so.

This leads to the final aspect of war's temporal boundaries. Just as beginnings are reflective of the power dynamics in international discourse—the war on terrorism only begins once a state declares it so—endings are equally the product of contested interpretation. Battles continue over the end point. In the modern imaginary with its bias towards state-state conflicts as the dominant conception of war, the idea of a war ending is a simple one. Instruments of surrender are signed and both sides agree that the conflict is over through mutual declaration. Rarely, if ever, in modern warfare is an actual war concluded through the annihilation of an actor in the conflict (Reiter 2003). But as is demonstrated by the increasing popular visibility of terrorist campaigns, civil conflicts, and insurgencies, the end of the war is a moment of interpretation. If citizens of their state choose to attack the adversary post-agreement it becomes a civil matter, not an indication that the war has restarted. In the case of conflicts where states are no longer the sole actors, such as terrorist campaigns, this is especially difficult because groups could take it upon themselves to continue the campaign even after agreements are in place that have 'ended' the conflict, as was the case in Northern Ireland, throwing wide-open the question of whether the conflict has truly 'ended' (Little 2014).

Implications for IR

Observing that temporal dynamics are at play in conceptions of war may possess some prima facia value, but what are the implications for future IR work? First, it emphasises that war is only one form of violence that exists in the international space. While this is not a particularly new argument as feminist theorists and peace studies scholars have identified the manner in which contemporary foreign policy devalues structural violence, what a temporal lens offers is another means of explaining why this is the case (Hutchings 2007, Blanchard 2003, Barash and Webel 2003). Temporal commitments regarding the understanding of war as an event with a definite beginning and end, duration and conclusion, contribute to the privileging of instantiations of violence like war and armed conflict over more diffuse, ongoing, structural forms of violence. Conceptions of war as a time-bound, discrete event make it easier to understand and distinguish as something that is important and as a problem that is potentially solvable (Cox 1981). Actions that are not temporally compressed or time-bound remain harder to appreciate or approach in the collective political imaginary.

The preceding analysis also reveals that war is not necessarily something that 'breaks out' or is 'sparked' by events, but rather a series of events and uses of force that get interpreted in the present and future present into one broader, linked event. Dominant discourses and narratives shape how we read these multiple disparate events. Given the role of sovereign states in constituting these discursive frames, it is unsurprising that war as a contemporary concept privileges state-centric notions. A narrative discursively 'makes sense' in the present what has happened in the past and allows the set of actions to be conceptualised as a war rather than something different and perhaps inevitable.

Approaching war with a temporal frame also reveals that war is a category of violence, not an objectively occurring idea. It is temporally imposed and constructed. The future and the past meet in complicated and important ways that serve to frame our understanding of it and sovereign state subjects inevitably play a large role in policing the boundaries of the event. The representation of the event also necessarily refers to that which is outside the present and therefore absent. Much like terrorist attacks operate as an idea in process, an event that is in a constant state of becoming, wars are similarly fluid in interpretation (Lundborg 2012). Legal scholars have attempted to identify the meaning of the term and can only agree on the idea that it involves violence and organised groups (O'Connell et al 2004). Critical War Studies as a school of thought has provided important insight into these understandings of war's ultimate ontology. Attention to temporality can benefit these moves and vice versa (Holmqvist 2013). In particular, if it is the case that the ontology of war is neither fixed nor objective, temporal relationships are intrinsic to these understandings. Wars are much like MacKenzie's idea of 'legislating' and the inevitable distinctions between the reified notion of war and its actual ontological foundation are distinctly about how we understand relationships across time.

A third implication that arises from turning a temporal lens on the concept of war is that it exposes the value of reflexivity specifically when approaching the concept of war. This operates at two levels. At the scholarly level, given the indeterminacy of any understanding of war itself, making comparisons becomes increasingly complicated. This necessitates a reflexive analysis of the manner in which IR treats war as a category of violence similar enough to be compared across vast periods of time and space. If war is a temporal entity that is contingent and indeterminate, however, then IR needs to ask more nuanced questions regarding its cross-temporal comparisons and its privileging of generalisability when it comes to this foundational concept. Models of IR inquiry like the Correlates of War project that use linear time to make discrete claims and mark relationships might need to be rethought, especially with respect to their conclusions. Insights based on the Thirty Years War or even Vietnam when brought into the present or future environment may be comparing apples and oranges.

This leads to the second level at which reflexivity comes in—asking the questions who the scholarship is for and what its purpose is? If IR scholarship is produced in the present and access to the present and near-future is the primary area we can influence as time-bound entities then it forces us as scholars of this discipline to take seriously the question of purpose and value in our actions (Berenskoetter 2011, Amoreux and Steele 2015). What is the value in studying counterterrorism, for instance, if we become sceptical of the objective accumulation of time-less political claims or truths (Krause and Williams 2004, Jackson 2015, Um and Pisoiu 2015)? If political claims, ideas, and knowledge are essentially time-bound, then the question of what value the claims have for the present and near-future become much more resonant and important. One can read the moves towards forecasting and mid-level theory as an implicit reaction to this issue (Mellers et al 2015, Dunn, Hanson, and Wight 2013). By narrowing their claims and/or taking an approach that actively theorises how their model applies to particular futures—rather than relying on an assumption that adding knowledge to the world will make prediction 'better', these approaches take on a more temporally limited vision that has the potential to better capture temporal complexity in our scholarly work. If part of the goal of studying war is to better understand it so as to limit the future suffering it causes, then reflexive analysis can do important work along these lines. Valuing the constitutive temporal dynamics at play is vital to this work.

Conclusion

To bring this back to the question of war in the present, it appears that the picture regarding what is going on between the United States and various terrorist groups and regimes in the Middle East remains murky. Declaring it a single war seems inappropriate given the multiplicity of adversaries, yet treating Afghanistan, Iraq, and Syria (as well as potentially Pakistan, Somalia, and Yemen) as completely separate conflicts seems equally problematic. What this chapter has sought to demonstrate, however, is that this problem is not unique to present-day politics, but is intrinsic to the concept. On the eve of what would become World War II, Carl Schmitt observed

> For several years now, in the most scattered parts of the Earth, sanguinary battles have been fought out while a consensus about the notion of and the term 'war' has been carefully avoided…It appears, what was always true, that the history of international law is the history of the notion of war… hence the notion of war becomes a problem the objective discussion of which is appropriate, in order to disperse the fog of deceptive fictions and allow the [emphasis added] *real* situation of *today's international law* to be recognized as such (Schmitt 1938, 1)

In some ways this captures how the currently unforeseeable future implicates the present but also the manner in which decisions in the present operate at the intersection of meaning in the past and future. What a temporal lens allows is the reintegration of the political imaginary of practitioners at the time. It functions as a tool to better interrogate and model ideas of temporality into predictions. It also reveals the temporal boundaries intrinsic to identifying particular acts of violence as occupying the privileged position of war. By revealing the temporal indeterminacy of the question Balibar asks, 'what is war?', attention to temporality opens space for similarly fundamental investigations to emerge via the de-naturalising of the conception of war itself.

References

Abbott, A. (2001) *Time matters: on Theory and Method*. Chicago: University of Chicago Press.

Amoreux, J. and Steele, B. (2015) *Reflexivity and International Relations.* Routledge Press.

Balibar, E. (2008) 'What s in a War? Politics as War, War as Politics.' *Ratio* 21(3): 365–387.

Barash, D. and Webel, C. (2013) *Peace and Conflict Studies*. SAGE Publications, Incorporated.

Becker, C. (1955) 'What Are Historical Facts?' *The Western Political Quarterly* 8(3):327-40.

Berenskoetter, F. (2011) 'Reclaiming the Vision Thing: Constructivists as Students of the Future.' *International Studies Quarterly* 55(3): 647-668.

Bialasiewicz, L., Campbell, D. and Elden S. (2007) 'Performing Security: The Imaginative Geographies of Current US Strategy.' *Political Geography* 26(4): 405–422.

Blanchard, E. (2003) 'Gender, International Relations, and the Development of Feminist Security Theory.' *Signs* 40(1).

Boon, K. and Lovelace, D.C. (2012) *The Changing Nature of War*. Vol. 127. Oxford: Oxford University Press.

Bourdieu, P. (1977) *Outline of a Theory of Practice*. Cambridge University Press'

Bush, G.W. (2001) Address to the Nation. September 11. Available online at: http://edition.cnn.com/2001/US/09/11/bush.speech.text/

Butler, J. (2006) *Precarious Life*. New York: Verso.

Butler, J. (2009) *Frames of War: When is Life Grievable?* New York: Verso.

Butler, J. (2012) 'Precarious Life, Vulnerability, and the Ethics of Cohabitation.'*The Journal of Speculative Philosophy* 26(2): 134-151

Buzan, B. and Lawson, G. (2013) 'The Global Transformation: The Nineteenth Century and the Making of Modern International Relations.' *International Studies Quarterly* 57(3): 620-634.

Clausewitz, C. von. (1984) *On War.* Translated and edited by Howard, M. and Paret, P. Princeton, NJ: Princeton University Press.

Cox, R. (1981) 'Social Forces, States and World orders: Beyond International Relations Theory.' *Millennium: Journal of International Studies* 10(2): 126–155.

Cronin, A. (2002/03) 'Behind the Curve: Globalisation and International Terrorism.'*International Security* 27(3): 30-58

Cronin, A. (2009) *How Terrorism Ends: Understanding the Decline and Demise of Terrorist Organizations.* Princeton: Princeton University Press. 7

Dettmer, J. (2015) 'The Age-Old Sunni-Shia War is Sucking America.' *Daily Beast.* March 30.

Dudziak, M. (2012) *War Time: An Idea, its History, Its Consequences.* New York: Oxford University Press

Dunne, T., Hansen, L. and Wight, C. (2013) 'The End of International Relations Theory?' *European Journal of International Relations* 19(3): 405-425.

Gleditsch, N. et al. (eds) (2013) 'The Decline of War.' *International Studies Review* 15(3): 396-419

Goodhand, J. (1999) 'From Wars to Complex Political Emergencies: Understanding Conflict and Peace-building in the New World Disorder.' *Third World Quarterly* 20(1): 13-26.

Harff, B. (2003). 'No Lessons Learned from the Holocaust? Assessing Risks of Genocide and Political Mass Murder Since 1955.' *American Political Science Review* 97(1): 57-73.

Hoffman, B. (2006) *Inside Terrorism*. New York: Columbia University Press (Revised and Expanded Edition)

Holmqvist, C. (2013) 'Undoing War: War Ontologies and the Materiality of Drone Warfare.' *Millennium-Journal of International Studies* 41(3): 535-552.

Hom, A. (2010) 'Hegemonic Metronome: The Ascendancy of Western Standard Time.' *Review of International Studies* 36(4): 1145-1170.

Hom, A. and Steele, B. (2010) 'Open Horizons: The Temporal Visions of Reflexive Realism.' *International Studies Review* 12(2): 271-300.

Hutchings, K. (2007) 'Happy Anniversary! Time and Critique in International Relations Theory.' *Review of International Studies* 33(1): 71-89.

Hutchings, K. (2008) *Time and World Politics: thinking the present*. Manchester University Press.

Hutchings, K. (2007) 'Feminist Ethics and Political Violence.' *International politics* 44(1): 90-106.

Jackson, P. and Nexon, D. (1999) 'Relations Before States: Substance, Process, and the Study of World Politics.' *European Journal of International Relations* 5(3): 291-332

Jackson, R. (2007) 'The Core Commitments of Critical Terrorism Studies.' *European Political Science* 6(3): 244-251

Jackson, R. (2015) 'The Epistemological Crisis of Counterterrorism.' *Critical Studies on Terrorism* 8(1): 33-54.

Jarvis, L. (2008) 'Times of Terror: Writing Temporality into the War on Terror.' *Critical Studies on Terrorism* 1(2): 245-262

Krause, K. and Williams, M. (1997) 'From Strategy to Security: Foundations of Critical Security Studies.' In: Krause, K. and Williams, M. (eds) *Critical Security Studies: Concepts and Cases*. University of Minnesota Press. 33-59.

Leatherby, L. (2015) 'Whatever Happened to the Debate over the Use of Force Against ISIS?' *NPR*. June 17. Available online at: http://www.npr.org/2015/06/17/415203016/whatever-happened-to-the-debate-over-use-of-force-against-isis

Little, A. (2014) *Enduring Conflict: Challenging the Signature of Peace and Democracy*. Bloomsbury.

Lipschutz, R. (1995) *On Security*. Columbia University Press.

Lundborg, T. (2012) *Politics of the Event: Time, Movement, Becoming*. Routledge.

McIntosh, C. (2015) 'Theory Across Time: International Relations' Search for Time-less Theory.' *International Theory* 7(3): 464-500.

Mearsheimer, J. (1994/1995) 'The False Promise of International Institutions.' *International Security* 19(3): 5-49.

Mellers, B. et al. (2015) 'Identifying and Cultivating Superforecasters as a Method of Improving Probabilistic Predictions.' *Perspectives on Psychological Science* 10(3): 267-281.

Musgrave, P. and Nexon, D. (2013) 'Singularity or Aberration? A Response to Buzan and Lawson.' *International Studies Quarterly* 57(3): 637-639.

The National Security Strategy of the United States of America (2015) Washington: White House.

O'Connell, M. (2005) 'Enhancing the Status of Non-State Actors Through a Global War on Terror?' *Columbia Journal of Transnational Law* 43: 435.

Power, S. (2001) 'Bystanders to Genocide.' *Atlantic Monthly* 288(2): 84-108.

Price, H. (1996) *Time's Arrow and Archimedes' Point: New Directions for the Physics of Time*. Oxford University Press.

Reiter, D. (2003) 'Exploring the Bargaining Model of War.' *Perspective on Politics* 1(1)

Reiter, D. (2009) *How Wars End.* Princeton University Press.

Sambanis, N. (2004) 'Using Case Studies to Expand Economic Models of Civil War.' *Perspectives on Politics* 2: 259-80.

Schmitt, C. (2003) *War/Non-War: A Dilemma.* Washington, DC: Plutarch.

Sewell, W. (1996) 'Three temporalities: Toward an eventful sociology.' *The historic turn in the human sciences*: 245-80.

Shultz, R. and Vogt, A. (2003) 'It's War! Fighting Post 11 September Global Terrorism Through a Doctrine of Preemption.' *Terrorism and Political Violence* 15(1): 1-30.

Solomon, T. (2013) 'Time and Subjectivity in World Politics.' *International Studies Quarterly* 58(4): 671-681.

Stevens, T. (2016) *Cyber Security and the Politics of Time.* Cambridge: Cambridge University Press.

Weberman, D. (1997) 'The Nonfixity of the Historical Past.' *The Review of Metaphysics:* 749-768.

Wendt, A. (1995): 'Constructing International Politics.' *International Security* 20(1): 71-81.

Wendt, A. (1999). *Social Theory of International Politics*. Cambridge: Cambridge University Press.

US National Security Strategy (2010) Washington: White House. Available online at: http://www.whitehouse.gov/sites/default/files/rss_viewer/national_security_strategy.pdf

9

Islam and the Politics of Temporality: The Case of ISIS

SHAHZAD BASHIR
STANFORD UNIVERSITY, USA

As human beings we are bound by time. While we can probably all agree that our experiences have temporal coordinates, beyond this, the matter gets more equivocal. What exactly is time? How does it function? In what direction does it flow (and what does it mean to say that it 'flows')? Abstract discussions of such matters abound in modern scholarship and are the province of physicists and philosophers. The understanding of time structures human cognition and response to the material world, which makes it a fundamental concern in the social sciences and the humanities. Time is an element within sociocultural imaginations that can vary greatly between contexts. Attending to temporality can be an advantageous venue for exploring complex and internally variegated topics such as the contemporary politics of Islam.

I should clarify that my concern in this chapter is only with the human experience of time. Following David Couzens Hoy, I will designate this 'temporality' in order to distinguish the matter from time perceived as an abstract universal (Hoy 2009: xiii). Temporality is implicated in human experiences and ways of thinking and has been the subject of extensive philosophical exploration (for example, Newton-Smith 1980). Within the study of international relations, temporal presumptions are the basis for value-laden categories that saturate popular media as well as specialised literature. The notion of progress, whether understood in liberal or Marxist frames, and the associated differentiation between economically developed and underdeveloped countries are premised on a temporal scale in which some people are seen as being ahead while others require catching up (Hutchings 2008). Such distinctions follow from an earlier anthropological positing of 'primitive' societies, understood as the past of Western societies despite being contemporaneous in the literal sense (Fabian 2014). In ordinary usage, Western temporality that dominates in the international sphere has become synonymous with abstract time. The elision has reified this temporality and its associated hierarchies, causing observers to understand its inner logic as a natural fact rather than the effect of a particular political regime that has been dominant in recent centuries (Hom 2010).

The conception of the international system as an object of study too rests on a particular temporality. It takes the current norms of Euro-American societies as the future of those outside this sphere, without reference to contingencies pertaining to sociocultural and historical differences (Inayatullah and Blaney 2004; Hobson 2012). Temporal presumptions underlie almost all understandings of international politics; occasional academic criticism of the hierarchical view of the world that results from this has had little effect on general public discourse in most parts of the world. While economic development is usually at the forefront of relative valuation, this always has sociocultural counterparts. Those who have less money are often thought to lag behind in moral, social, and cultural sophistication as well. In this way, Western temporal benchmarks that operate in the background of analytical paradigms tend to predetermine how the non-Western world gets portrayed in popular media and scholarship.

Islam between Medievality and Reformation

Temporal markers are a staple in present-day descriptions of the politics of Islam and Muslims, although this topic has not received extensive analytical attention. Consider, for example, the frequency with which the term 'medieval' is used to discuss contemporary individuals and groups claiming to act in the name of Islam.[38] From the 1980s, influential authors have described anti-establishment movements claiming Islamic identities as cases of amalgamation between medieval theology and modern politics (for example, Sivan 1985). In doing so, scholars, and the journalists to follow them, have taken the rhetoric of the groups they were attempting to understand at face value rather than seeing them as creative modern readers of selected premodern works. Western scholars were predisposed to this perspective because of the governing logic of orientalism, the academic and popular discourse that has framed Muslims as quintessential 'others' since the nineteenth century. In this view, Islam is seen to have an ahistorical essence that became settled at a point in the 'medieval' period. This essence has then

---

[38]   The term medieval is commonly used also to refer to Islamic contexts coeval with the period identified as the European Middle Ages. Such usage presents its own set of conceptual problems that are beyond the scope of the present discussion.

imprisoned Islam in some past time, making Muslims beholden to 'tradition' in a way that is not true for other religious communities. This perspective remains influential today: prominent scholars continue to argue for Muslims' exceptionality as political actors on the basis of claiming features supposedly inherent in Islam irrespective of context (for example, Cook 2014). As I have argued elsewhere, this is a theological view that has come to be seen as 'secular' history due to the way Western scholars have privileged certain Islamic discourses about the past as repositories of an essential Islam (Bashir 2014). I believe this academic framing needs a wholesale corrective. An alternative approach, more compelling to me, is to see Muslims as active agents who have, through history, created multiple pasts to serve intellectual and sociopolitical interests perceived as being relevant for their present situations.

The use of the term 'medieval' for contemporary Muslims derives from a seemingly self-evident, valorised temporal scale that has its origins in the purported historical trajectories of European societies. The term's efficacy rests on two steps. First, the adjective medieval is marked as a negative past, from which Europe and its socio-intellectual descendants are supposed to have progressed via development over time. As scholars of European history have emphasised, medievality is a modern intellectual construct whose chief purpose is to posit the superiority of the modern age (Symes 2011). And second, when contemporary Muslims actors are referred to as medieval, they are relegated to Europe's negatively valued past. The result should look incongruous: people adept at the use of electronic media and advanced mechanical weapons, operating in Asia and Africa, are being imaged in a European world associated with manuscripts and swords. The ludicrousness of the juxtaposition misses the eye because medieval here is a marker of moral disapprobation rather than a description of the peoples in question. The term's deployment accomplishes the double task of distancing 'us' from surpassed ancestors and unworthy contemporaries. Those who use the term for Islamic groups certainly mean it to be negative. Placing this judgment within a seemingly self-evident time scale reifies the commentators' own temporality and renders it above critical assessment. The temporal argument enfolded within the use of the term medieval creates the illusion that the judgment on offer is an objective fact rather than being an expression of particular intellectual and sociopolitical interests.

While medieval implies adjudication on the basis of a scale of general cultural development, the view that Islam either needs, or is experiencing, a 'reformation' transfers the same temporal logic to a specifically religious arena. The idea of a totalising reformation, which is different from general piecemeal change and reform, has been raised by Muslims and non-Muslims alike, belonging to many different places on the contemporary political spectrum (an-Naim 1996; Browers and Kurzman 2004; Ali 2015, and many others). The obvious point of reference here is the history of Christianity, which is being universalised and made the supposed timeline for Islam. Just as the Protestant Reformation purportedly redeemed Europe from Catholic Christianity, the helping hand of modern European ideas is supposed to make Muslims into secular modern peoples (Mahmood 2006). The effect is to negate the possibility that the religion of Muslims may have a history independent of Christianity, connected not to views of Europe but to contingencies pertaining to the histories of their own communities. This kind of elision between Christianity and Islam is foundational also to the category religion, which evolved out of Christian ideas and has universalised the application of Christian patterns across non-Christian contexts (Masuzawa 2005).

To call certain contemporary Muslims medieval, and to suggest that Islam needs a reformation, are matters that index unstated temporal presumptions and their associated moral verdicts. The use of such terms masks cultural, political, and military agendas and is helpful for generating public fervour for the actions of certain states. However, such usage should have little analytical purchase when it comes to understanding the actors to whom the terms are applied. In fact, judging Muslims and others according to historical trajectories of Western societies is a widespread fallacy that has been the cause of obfuscation rather than analytical advancement. To correct this situation, it is necessary to consider the particulars of temporalities embedded within non-European discourses. This has the benefit of relativising Euro-American perspectives while also providing more meaningful access to human experience in a broader, more universal frame.

I should emphasise that my purpose here is not to posit a binary opposition between Western and Islamic temporalities. Ideas and technological developments that originated in the West have had a profound effect on how Muslims today understand their own pasts. And we can also show that Muslim and other temporalities matter for understanding the West. But crosscutting impact of this nature cannot be seen as preordained according to the way Euro-American societies may have developed. To assess Muslim temporalities requires analysis of data emanating directly from the subjects in question. Recent discussions in the philosophy of history have shown that Western societies have been (and are) host to many different understandings of temporality (Jordheim 2014). The same needs to be presumed for Islamic contexts: analysis should proceed from the fact that Muslim understandings of the experience of time are multiple and changeable. When it comes to temporality, neither the 'West' nor 'Islam' are hermetically sealed entities. Both words reference internally variegated fields, encompassing description as well as rhetoric, in which temporalities are critical elements within the evolution of ideas and practices. To concentrate on Islamic temporalities is, therefore, the opposite of the effort to specify the exclusive essence of Islam as sought in orientalist scholarship or Islamic theological discourses.

The Rise of ISIS

In contemporary international affairs, the group that has come to be known as ISIS (Islamic State in Iraq and al-Sham) or Daesh (ad-Dawla al-Islāmiyya fī l-Irāq wa-sh-Shām) is a particularly vivid case for the politics of temporality and Islam. Here is an entity that has frequently been called medieval since its appearance. Moreover, the group's propaganda actively promotes the understanding that it is a kind of re-enlivening of an Islamic past radically at odds with orthodoxies identified with modernity. We can see this prominently in its declaration of the worldwide caliphate on 29 June 2014, recalling a religiopolitical formation with origins in the seventh century CE (Poirson and Oprisko 2014; al-Rasheed, Kersten and Shterin 2012). Its self-proclaimed founding leader is known by the *kunya* (part of an Arab name) Abu Bakr, referring to the name of the first successor to Prophet Muhammad. The choice of the name implies an erasure of time between the seventh and the twenty-first centuries CE.

Reports issued by the United Nations indicate that populations of the territory under the control of ISIS have been subjected to policies that the group claims were the military-political norms of the early Islamic period. This has involved widespread summary use of force against groups identified as Yezidis, Christians, Kaka'is, Kurds, Mandeans, the Shi'a, and the Turkmen. The treatment has reportedly included killing, torture, and forced conversion of men and boys, and the distribution of 'unbelieving' women and girls as sexually permissible 'spoils of war' among male ISIS fighters (UN Report 2015).[39] For the case of sexual slavery, the group has issued a five-page pamphlet that mimics traditional Islamic legal discourse in a highly reductive way (Anonymous 2014). It provides absolute opinions in the form of 34 questions and answers that justify the group's practices. To date, examinations of statements and acts attributed to ISIS have attempted to place them in the timeline of practices known from other Islamic contexts. Such discussions privilege an Islamic tradition continuous over time and adjudicate whether, and how, ISIS corresponds with earlier perspectives (McCants et. al. 2015). Inasmuch as ISIS ideologues do selectively cite Islamic sources such as the Quran and hadith, should we take their rhetoric of the unmediated return to an earlier Islamic era on face value?

---

[39] I have had no direct access to proponents of ISIS or the territory controlled by the group. The scope of my analysis here is limited to what ISIS itself has disseminated and what has been reported widely in news media.

I argue that we should be sceptical regarding the claims made by ISIS. However, my reason for saying this is not that I believe Islam to have essential characteristics that are absent in the group. In historical perspective, Islam cannot have an essence since a religion has to be understood as that which its proponents think and do in all its variety. Whenever an essence is invoked in a religious context, a theological rather than sociohistorical claim is coming into play. Islam should be seen as a perennially changeable affair described through reference to historically locatable ideas and practices. I suggest that ISIS operates on the basis of temporal processes and understandings that make its politics very much a contemporary affair. Rather than mapping continuities or discontinuities with a tradition, proponents of ISIS should be seen as producers of a discourse that deploys its temporality for sociopolitical ends. Temporal understandings and practices both structure its observable perspective and are available for manipulation for propagandistic purposes.

An Electronic Caliphate at War

ISIS has weaponised time. While ISIS is neither the first nor the sole entity to do so, the movement is the most prominent Islamic case of the phenomenon present on the world stage at the moment. It is a warring faction operating in a region that has been subject to a foreign invasion and endemic political insecurity for more than a decade. Temporal elements projected in its propaganda are parts of the arsenal it deploys through present-day ways and means. In the movement's ideological self-presentation, the world is divided between believers and non-believers, 'us' and 'them', on the basis of claiming absolute knowledge of a distant past. Through a kind of temporal telescoping, affinities and antagonisms sedimented over fourteen centuries are stripped of their historical contexts and are purveyed as absolute religious truths. This has resulted in practices that are justified through historical precedent but without allowing for contextual contingency or change over time. This is a temporality with a very shallow horizon whose very sparseness and crudity is presented as the guarantee of its veracity.

ISIS deploys the past as a weapon primarily through the internet, a medium that has become available widely only during the past quarter of a century. The internet provides a spatiotemporal field quite different from earlier media such as manuscript culture, print, and broadcasting. It allows for instantaneous dissemination of a universalising ideology, such as a caliphal state that is meant to administer the affairs of Muslims on a global scale. While the idea of the caliphate has appealed to Muslim ruling elites at various points in history, its deployment by ISIS is unique and is made possible by the availability of the internet. To the best of my knowledge, no other declaration of the caliphate in Islamic history has ever brought adherents from many corners of the world into the lap of a political movement within a very short period. Here, the internet as a medium is not simply a novel way to distribute propaganda. Rather, the movement's highly charged, yet minimalist message is engineered to maximise the potential inherent in the means of communication. Thanks to the internet, ISIS has a global audience that responds to its messages instantaneously, whether positively or negatively. The movement's ideological output is keyed to this temporal advantage; the partial success of the strategy is visible in stories about people trekking from all over the world to join the group.

The internet has created a new kind of 'Islamic world' where it is possible to enact and maintain a globally interconnected Islamic identity irrespective of geographical distances. If al-Qaeda, deterritorialised after 2001, was an incipient stage of this development (Devji 2005), ISIS presents a more mature and administratively adept case of the phenomenon. The internet allows ISIS to disseminate its propaganda speedily, in high, uninterruptible frequency. The effects of this ideological projection vary based on the social context of the point of reception. Among likely sympathisers, the minimalism of the propaganda allows for it to be read variably, creating multiple paths through which people can join the movement. The resulting 'community' is one part virtual, spread around the world, and another part localised to the regions in Iraq and Syria where the movement holds political power. The movement's virtual and real sides are mutually reinforcing: electronic messaging creates the conditions for people to join the group while the expansion in ranks leads to enlarging its presence on the internet. For the movement's detractors, the shallowness of the propaganda makes it jejune, its attraction for some seemingly inexplicable. However, the minimalist message, and the medium through which this is delivered, are keyed to the low common denominator that is presumed to be dispersed in a global audience. The presence of ISIS propaganda on Euro-American television, computer, and mobile phone screens is part of the same process that carries Western ideologies in the form of cultural products to other parts of the world. In the enactment of an electronic caliphate, contraction of time and space resulting from recent proliferation of electronic communication may have turned out to have startlingly unpredictable consequences.

Archaeology of an Emblem

A consideration of the imagery found on the ISIS flag highlights a different but equally significant vector of temporality. The flag predates the prominence of ISIS and has been used by numerous other groups that have had similar political and cultural agendas. It contains two Arabic texts: on top, the phrase 'there is no god but God' in white with black background, and on the bottom, 'God, Messenger, Muhammad' in black inside a white roundel. In themselves, these words are unremarkable in an Islamic context. They indicate the first part of the affirmation of faith and the declaration of Muhammad's prophecy. The flag's distinctive 'message' lies in its colour (black has been associated with revolutionary movements) and the crudeness of the script that is meant to signify closeness to Islamic origins. The flag of the Kingdom of Saudi Arabia contains substantially the same text, in elegant script in white placed on green background. The use of this flag by ISIS and other groups likely reflects contestation over the sphere claimed by Saudi Arabia as the state that controls the lands of Islam's origins and regularly portrays itself as the guardian of Sunni Islamic values.[40]

When considering temporality, the most interesting aspect of the ISIS flag is the crudeness of the script and the placement of the three words in a roundel. The latter is thought to represent the impression of a seal used by the Prophet Muhammad. No reliable data is currently available on the exact source of the seal image, which is an important point of information in itself. What matters is the image, differentiated from the flowing cursive of the flag of Saudi Arabia and asserting authority through a generalised visual appeal to antiquity. The purported seal, whatever its provenance, may have acted as a relic, an object of religious power utilised by elite patrons. What we have here, instead, is an image without the original, whose potency resides in its endless reproduction through mechanical and electronic means. The image's lack of symbolic specificity invites varied interpretations, allowing it to be a unifying factor without requiring ideological depth or substantive conformity of purpose.[41]

The image contained in the flag signifies age through its form, recalling objects usually encountered in modern museums. Such institutions embody a

---

[40]   The Saudi state is itself a valuable case for thinking about Islamic temporalities. It involves a blending of imperatives pertaining to a dynastic polity, a modern nation-state, and hyperterritorial championing of 'Islamic' causes around the world (al-Rasheed, 2006; Lecroix, 2011). Examining the details of the issue can aid in getting beyond simplistic evocations of 'Wahhabism' that still proliferate in discussions of the Kingdom.

[41]   The use of the seal image by ISIS recalls Walter Benjamin's famous 1936 essay on art in the age of its mechanical reproducibility (2002): '[T]echnology of reproduction detaches the reproduced object from the sphere of tradition. By replicating the work many times over, it substitutes a mass existence for a unique existence. And in permitting the reproduction to reach the recipient in his or her own situation, it actualises that which is reproduced' (104). While Benjamin's reference points were photography and film, the internet has further radicalised the issue. The electronic image is the central emblem of ISIS, available globally as an ideologically meaningful object without the need of an elaborate undergirding tradition.

particular ideological perspective on the past connected to Western epistemologies that have by now globalised (Karp 2006). Again, the contrast with the text on the flag of Saudi Arabia is instructive and reflects a mass objectification of the past that is of quite recent provenance. The image indicates an archaising appeal to authority, akin to the way modern states use images of objects unearthed and preserved by archaeologists as national symbols. The region in which ISIS predominates has received extensive archaeological attention since the nineteenth century. Iraqi and Syrian nationalisms have strong archaeological components (Bernhardsson 2005), reflected in the proliferation of images on common use objects such as currency and stamps in both countries. These usages project *longue durée* genealogies for the nation, displaced from the messy circumstances of the present, yet fully tangible in the form of museum objects and buildings of the past that can be visited in the course of nationalist pilgrimages.

The symbol of ISIS is the image of an old object—and not simply a text—that similarly attempts to eradicate sociohistorical complexities through appeal to pristine primordiality. The movement's well publicised antipathy towards certain antiquities, exhibited in the selling or destruction of objects, is better understood as the effort to eliminate competition rather than the exercise of some age old Islamic iconoclasm. As in the case of Taliban destruction of the Buddhas in Bamiyan, Afghanistan, in 2001, locally immediate material and ideological conditions are far more significant here than religious ideas often portrayed as essential to Islamic belief. For both ISIS today and the Taliban earlier, the occlusion of these factors in most Western media coverage reflects the power of unconscious presumptions regarding Muslims and the past (Elias 2007; Bernbeck 2010).

The End of Time

Dabiq is an official online magazine published by ISIS in multiple languages since Ramadan 1435 (July 2014). Between the first and the tenth issue (July 2015), it has grown in size from approximately 50 to 75 pages. The magazine's contents are diverse: religiopolitical proclamations; reports about the heroics of ISIS fighters, accompanied by male faces, smiling, serious, or covered with scarves; condemnations of the movement's enemies, identified in religious terms pertaining to non-Muslims or Muslims declared to be wayward; appeals to readers to join the movement; and, since issue 7 (January-February 2015), a section especially for women. Matching the videos the group is known for posting on the internet, the magazine sometimes highlights gruesome images of dead bodies, whether of its own 'martyrs' or of enemies.

The magazine has a glossy quality, including frequent artful deployment of the distinctive images on the flag I have discussed above. The writing is generally simple and declarative, with little pretence towards analysis. It does, nevertheless, have a highly cultivated quality so that the writers' own speech appears very much the way Quran and hadith reports usually sound in English. This presumably carries to versions of the magazine in other languages too, although I do not have access to such texts to provide a concrete judgment. Overall, the magazine's text has an archaising tone that matches the affect of the movement's electronic messaging and its flag.

The magazine's first issue states that its name is 'taken from the area named Dabiq in the northern countryside of Halab (Aleppo) in Sham. This place was mentioned in a hadith describing some of the events of the Malahim (what is sometimes referred to as Armageddon in English). One of the greatest battles between the Muslims and the crusaders will take place near Dabiq' (p. 4). The explanation is interesting in terms of the movement's ordering of time: the name is anchored in an expected future justified through a past that has been flattened into a perennially existing binary between self and other. This is, I would suggest, a past constructed in the image of the future rather than the other way around. The religious battle presaged here carries the flavour of a coming apocalypse in the material sphere. This sense is confirmed in the magazine's second issue (Ramadan 1435/July 2014), whose cover carries the partial image of a large wooden boat accompanied by the statement 'It's either the Islamic State or the Flood'. The larger story connected to this pronouncement (pp. 5-7) portrays the group as the world's last potential saviour prior to the advent of the kind of divine punishment that is associated with Noah in biblical/Quranic accounts. The pattern of thematising the future through reference to iconic events placed in the past pertains to all the issues that have appeared to date.

The magazine's overall coverage of topics as well as the tone of the writing indicate what I would call a profoundly 'presentist' orientation. That is, categories and concerns of the movement's situation in the immediate present stand behind all its projections regarding the past. Given this attribute, it seems quite irrelevant to use terms such as 'modern' and 'medieval' to describe the group. The most detailed treatments are devoted to stories picked up from news media, and the movement's writers seemingly scour the internet for reports and commentary about themselves. Even when engaging matters beyond present interest, the magazine's primary drive is towards proselytization aimed at immediate recruitment. For example, in issue number 10 (July 2015), an article published under the heading 'From the Pages of History' is ostensibly concerned with showing that the month of Ramadan is an auspicious time for military action. This is argued through brief, decontextualised presentations of Muhammad's warfare as reported in selected early Islamic sources, followed by direct address to the reader to take up arms in support of ISIS. The story is accompanied by two three-quarter page photographs showing men with faces covered or barely visible, riding horses with swords in hand (26-28). These are all clearly theatricalised performances that are projecting an eternalised past usable for the immediate present.

In recent media coverage of ISIS, Graeme Wood's article 'What ISIS Really Wants' in the 25 March 2015 issue of *Atlantic Monthly* has received considerable attention. Observing ISIS self-proclamations, Wood suggests that we take the movement seriously as an apocalyptic group that derives its programme from a 'medieval' Islam that can be tracked in literary sources. He is critical of those who argue that ISIS should be regarded as un-Islamic since the apocalyptic rhetoric has clear Islamic origins. Responding to Wood's position, critics such as Caner Dagli have argued that ISIS ideologues are manifestly ignorant of the complexities of Islamic religious history and literary sources. This should, in the view of such commentators, disqualify the movement from being taken seriously as an Islamic perspective. Both these understandings are premised on the presumption that the Islamic past is something 'out there' that can be uncovered well or badly on the basis of universally applicable criteria. Both judge ISIS actors by evaluating media performances as unmediated windows onto the beliefs of the movement's proponents.

Wood's perspective is overly simplistic. He interprets rhetoric as straightforward reality, reflecting a problem that has been pervasive in media coverage of ISIS (Doostdar 2014). He is beguiled by the supposedly self-evident logic of the apocalypticism articulated by ISIS and fails to see that such ideas are inseparable from the historical contexts of which they form a part. Messianism and apocalypticism have a long history in Islamic thought and their political appeal has waxed and waned based on circumstances. Such ideas should scarcely be seen as a supra-historical 'true' Islam that acts as the engine behind movements such as ISIS. Wood is surprised by the fact that supporters of ISIS spout age-old texts at the same time as they act in modern cosmopolitan ways. This observation makes eminent sense when seen in conjunction with the fact that the movement's presentist orientation drives its severely minimalist vision of the past. For ISIS, the past is a very short script, easily placed within an overwhelmingly larger concern with an immediate present in which technological innovation and deployment are entirely normative. The views of Dagli and other similar critics of Wood are, on the other hand, expressions of Islamic theology in competition with ISIS. Their purpose is to differentiate between right and wrong Islam, on the basis of self-consciously Islamic ideological and political commitments. Sidestepping both these perspectives, I have suggested that it is more analytically satisfying to excavate temporal features that are crucial to ISIS actions and propagandistic self-assertion.

Conclusion

My attempt to understand ISIS is invested neither in channelling the movement nor in competing with it on the basis of alternative religious truth claims. ISIS's primary medium for self-expression (the internet), its all-pervasive symbol (the flag), and its most elaborate propaganda instrument (the magazine) share a host of features pertaining to temporality. Its intellectually sparse yet absolutist propaganda utilises its vision of the Islamic past as a weapon against its detractors, Muslim as well as non-Muslim. The movement's projections regarding the past are generated with attention to what works well in the context of communication over the internet. It has leveraged twenty-first century technology to create a shallow temporality in which its own purportedly eternal truth is pitted against the falsehood represented by all opponents. The movement has self-consciously created a visual brand, focused on the past, which is promulgated through wide circulation of the symbolism present on its flag. It manipulates the media (including, most prominently, Western news sources) seemingly with the aim of blurring the boundary between its global electronic presence and its control over specific territory in Iraq and Syria. The spatiotemporal collapse contained within this manoeuvre is of a piece with its severely impoverished understanding of the Islamic past and future. Both tactics serve the purpose of creating a political image whose potency is said to reside in its simplicity. I hope to have shown that the movement's minimalist outward appearance overlays a host of complexities pertaining to the contemporary political and media environment. To continue to understand the movement through surface readings of the logic presented in its rhetoric may be a case of aiding in its success as a player on the world stage.

I have attempted to show that temporality is fundamentally a political matter that can be analysed through attention to its deployment in rhetoric and material production. ISIS is not unique when it comes to the significance of temporality in the creation of political programmes. Expanding the field of vision beyond the movement, issues pertaining to the past have been a constant factor in Islamic thought from the seventh century to the present (Bashir, forthcoming). In the contemporary setting, the kind of analysis I have attempted can be extended to other contenders, wherever they may fall on the political spectrum. Whether categorised as progressive, liberal, modernist, feminist, traditionalist, Islamist, radical, or extremist, contemporary Muslim sociopolitical actors create particular visions of Islamic pasts serviceable for varying ends. This is so especially in cases where a movement wishes the future to be different from what it perceives as the present. Moreover, all temporalities are products of particular circumstances and none predetermines events and experiences that are yet to materialise. Whatever the wishes of proponents of ISIS and other movements, actual futures will follow new contingencies that will, in turn, generate their own versions of pasts and futures. Herein lies part of the special complexity of the topic at hand: while temporalities can be objectified for analytical purposes, all explanatory narratives we can produce will contain their own debts to temporal orders whether conscious or unconscious.

This publication was made possible in part by a grant from Carnegie Corporation of New York. The statements made and views expressed are solely the responsibility of the author.

References

Ali, A. H. (2015) *Heretic: Why Islam Needs a Reformation Now*. New York: Harper.

Anonymous (2014) *Su'āl wa-jawāb fī s-sabī wa-r-riqāb*. Maktaba al-Himma, Dīwān al-Buūth wa-l-iftā'.
Bashir, S. (2014) 'On Islamic Time: Rethinking Chronology in the Historiography of Muslim Societies.' *History and Theory* 53 (4): 464-519.

Bashir, S. (forthcoming) 'The Many Spirits of the Islamic Past'. In: Cornell, V. and Lawrence, B. B. (eds) *The Wiley-Blackwell Companion to Islamic Spirituality*. Chichester, UK: Wiley-Blackwell.

Benjamin, W. (2002) 'The Work of Art in the Age of Its Technological Reproducibility: Second Version.' In: Eiland, H., and Jennings M. W. (eds) *Walter Benjamin: Selected Writings Volume 3 1935-1938*. Cambridge: Harvard University Press.

Bernbeck, R. (2010) 'Heritage Politics: Learning From Mullah Omar?' In: Boytner, R., Dodd, L. S., and Parker, B. (eds) *Controlling the Past: Owning the Future: The Political Uses of Archaeology in the Middle East*. Tucson: University of Arizona Press.

Bernhardsson, M. (2005) *Reclaiming a Plundered Past: Archaeology and Nation Building in Modern Iraq*. Austin: University of Texas Press.

Browers, M. and Kurzman C. (eds) (2004) *An Islamic Reformation?* Lanham, Maryland: Lexington Books.

Cook, M. (2014) *Ancient Religions, Modern Politics: The Islamic Case in Comparative Perspective.* Princeton: Princeton University Press.

Dabiq. Issues 1-10. Available online at: https://en.wikipedia.org/wiki/Dabiq_(magazine)

Dagli, C. (2015) 'The Phony Islam of ISIS.' *The Atlantic.* Available online at: http://www.theatlantic.com/international/archive/2015/02/what-muslims-really-want-isis-atlantic/386156/

Devji, F. (2005) *Landscapes of the Jihad: Militancy, Morality, Modernity.* Ithaca: Cornell University Press.

Doostdar, A. (2014) 'How Not to Understand ISIS.' Available online at: https://divinity.uchicago.edu/sightings/how-not-understand-isis-alireza-doostdar

Elias, J. (2007) '(Un)making Idolatry: From Mecca to Bamiyan.' *Future Anterior: Journal of Historic Preservation* 4(2): 2-29.

Fabian, J. (2014). *Time and the Other: How Anthropology Makes its Object.* Reprint Edition. New York: Columbia University Press.

Hobson, J. M. (2012) *The Eurocentric Conception of World Politics: Western International Theory, 1760-2010.* Cambridge: Cambridge University Press.

Hom, A. R. 'Hegemonic Metronome: The Ascendancy of Western Standard Time.' *Review of International Studies* 36(4): 1145-1170.

Hoy, D. C. (2009) *The Times of Our Lives: A Critical History of Temporality.* Cambridge, Massachusetts: The MIT Press.

Hutchings, K. (2008) *Time and World Politics: Thinking the Present.* Manchester: University of Manchester Press.

Inayatullah, N. and Blaney, D. (2004) *International Relations and the Problem of Difference.* New York and London: Routledge.

Jordheim, H. (2014) 'Introduction: Multiple Times and the Work of Synchronization.' *History and Theory* 53(4): 498-518.

Karp, I. et al. (eds) (2006) *Museum Frictions: Public Cultures/Global Transformations*. Durham: Duke University Press.

Lacroix, S. (2011) *Awakening Islam: The Politics of Religious Dissent in Contemporary Saudi Arabia*. Cambridge, Massachusetts: Harvard University Press.

Mahmood, S. (2006) 'Secularism, Hermeneutics, Empire: The Politics of Islamic Reformation.' *Public Culture* 18(2): 323-347

Masuzawa, T. (2005) *The Invention of World* Religions. Chicago: University of Chicago Press.

McCants, W. et al. (2015) *How Does ISIS Approach Islamic Scripture (Parts 1-4)*. Markaz: Middle East Politics and Policy. Available online at: http://www.brookings.edu/blogs/markaz/posts/2015/03/24-isis-approach-to-islamic-scripture-part-one-bunzel

An-Naim, A. A. (1996) *Toward an Islamic Reformation: Civil Liberties, Human Rights, and International Law*. Syracuse: Syracuse University Press.

Newton-Smith, W. H. (1980) *The Structure of Time*. London: Routledge & Kegan Paul.

Poirson, T. and Oprisko, R. (eds) (2014) *Caliphates and Islamic Global Politics*. Bristol: E-International Relations.

Al-Rasheed, M. (2006) *Contesting the Saudi State: Islamic Voices from a New Generation*. Cambridge: Cambridge University Press.

Al-Rasheed, M., Kersten, C. and Shterin, M. (eds) (2012) *Demystifying the Caliphate: Historical Memory and Contemporary Contexts*. Oxford: Oxford University Press.

Sivan, E. (1985) *Radical Islam: Medieval Theology and Modern Politics*. New Haven: Yale University Press.

Symes, C. (2011) 'When We Talk About Modernity.' *American Historical Review* 116(3): 715-726.

Wood, G. (2015) 'What ISIS Really Wants.' Available online at: http://www.theatlantic.com/magazine/archive/2015/03/what-isis-really-wants/384980/

10

Disrupting the 'Conditional Selfhood' of Threat Construction

KATHRYN MARIE FISHER
NATIONAL DEFENSE UNIVERSITY'S COLLEGE OF
INTERNATIONAL SECURITY AFFAIRS, USA

The Need to Disrupt Identity

In thinking about time, temporality, and global politics, I propose that we think about conditional (non)belonging in the context of identity, insecurity, and counterterrorism. If we are to better combat counterproductive consequences of othering that increase *in*security, we must critically investigate threat labels and the meanings and policies that they legitimise through exclusionary us/them boundary-drawing. This requires that we dislocate status quo time horizons and associated identity assumptions, and that we prioritise empathy, imagination, and analytical risk-taking. It is through this disruption of time, being, and (non)belonging, I would argue, that we have the best chance of achieving effective *and* ethical security strategies.

This discussion contributes to ongoing conversations destabilising misperceptions of (non)belonging. This destabilisation helps us to see counterproductive effects of security practice, however unintentional these effects and their resultant insecurities may be. The consequences of dehumanisation stemming from narratives of threat construction affect individuals with no relation to violence, a consequence that is enabled and worsened by generalisations of collectives and characteristics as both 'risky' and 'at risk' (Heath-Kelly 2012).

In this sense some actors are problematically positioned along what I consider a conditional state of belonging: as part of the self, but as somehow always on the cusp of otherness and insecurity. This is ineffective with respect to security objectives, and counterproductive with respect to human rights and social justice. Identity labels are not objective signposts but ambiguous signifiers, the allocations of which too often depend upon exclusionary us/them constructions. It is our responsibility to disrupt the presumed parsimony of such representations to fight the experiences of insecurity that they enable.

How can we think about identity disruption?

There are a number of cases through which we could situate discussion on conditional belonging such as the Mediterranean migrant crisis and exclusionary discourse in Europe, xenophobic debates on immigration in US presidential campaigns, and ongoing racism in society-police relations across the US. There are, indeed, too many cases. In these examples certain actors are positioned on the (often literal) borders of full belonging, even as they do not pose a threat to security. We must ask how a temporality of conditional selfhood is formed and exacerbated by processes of boundary construction separating 'us' and 'them'. It is impossible to develop effective security strategy if we do not consider how *in*securities are enflamed by identity (mis) perceptions associated with security discourse and practice.[42]

For this discussion, I would like us to critically engage with the idea of homegrown terrorism, by which I mean viewing threat construction as a result of social and political practice rather than an expression of objective truth. In this sense the focus is on how we conceptualise the very idea of a 'homegrown other' as well as how we observe articulations of homegrown in discourse. It is an admittedly messy focus given the tensions arising from examining both idea and articulation in such a short piece.[43] However, it is hoped that this will still enable a useful starting point and introductory conversation. When counterterrorism relies so heavily on the liminal space between an imagined act of violence and an actual act of violence, the very idea of identifying a homegrown other in preventive security practice relies upon an uncertain and consequential temporal plane of targeting.[44] A critical approach does not refute the possibility of violence from some identified as

---

[42]   "Discourse" here refers to language from official reports and websites as well as media and academic discussion. It is thus methodologically broad, but serves the intent of this introductory piece.

[43]   This tension demands a much fuller response than given here.

[44]   With many thanks to an anonymous reviewer on the liminal spaces of counterterrorism.

'homegrown', but does destabilise any notion of 'homegrown' as having self-evident meaning.

This helps us consider temporal dynamics of conditional (non)belonging in how processes of boundary construction often rely upon an association of terror with already marginalised groups: those seen as not having been part of the self for 'long enough', such as immigrant and Muslim populations. It is not that this is the only way that homegrown is constructed. There is no single definition of homegrown and there have been instances where 'homegrown' is applied to actors unrelated to Islamic extremism. Importantly, the Assistant Attorney General for National Security at the U.S. Department of Justice very recently said 'Homegrown violent extremists can be motivated by any viewpoint on the full spectrum of hate — anti-government views, racism, bigotry, anarchy and other despicable beliefs', that 'no single ideology governs' (Williams 2015). Thus even as there was no reference to terrorism, some articulations of homegrown include white supremacist and anti-government acts in addition to jihadist extremism.

At the same time, despite this plurality of meanings, a predominant connotation seems to stem from assumptions of distance and difference as danger in relation to homegrown and Muslim groups. Further empirical research is necessary, but as a starting point this chapter engages two claims: One that such research needs to be pursued, and two, that an identity-conceptualisation of conditional (non)belonging is one way to pursue it. Through this lens we can see how overcoming insecurity requires overcoming the damaging application of outsider status to those positioned as somehow not *yet* a part of 'us', as always encompassing some degree of otherness. Given local and international narratives that the West is at war with Islam, in addition to insecurity for those who are part of the self and have no relation to violence, we must ask how 'security' can create insecurity, regardless of intent.

In considering the dehumanising tropes that ambiguous and consequential threat articulations often employ, the cost of using homegrown labels may outweigh the benefits. By breaking down labels we may better empathise with (without speaking for) victims of insecurity, combating politics of non-belonging without falling into parochial, racialised, and/or orientalist[45] discourse (Said 1979; Biswas 2004). This breaking down is necessary because in 'doing security' discursive representations and material practice often (mistakenly) conflate groups of (often minority) individuals as collectively under-civilised and under-developed (Hindess 2007). When this

---

[45]   For an important study of homegrown as the 'orientalised insider' see Chuang and Romer (2013).

misperceived lack of civilised-being is linked to terrorism, assumptions of under-development are even more consequential. For example, in processes of externalisation, including specific counterterrorism law exclusion orders that enable state powers to send those under suspicion of terrorism 'back to' another place (Finighan 2014; Fisher 2015). In this way distancing self and other depends upon and reinforces mistaken associations of *difference* as *danger*. The homegrown identifier is one example with which we may better analyse the significance of such distancing.

Homegrown as Problematic Identifier

In 2013 James B. Comey, head of the FBI, positioned the 'emergence of home-grown violent extremists in the United States' (Horwitz 2013) as the threat that he wakes up to every morning and goes to bed with every night, and in 2013 President Barack Obama identified 'homegrown extremists' as 'the future of terrorism' (2013). Given this as well as the known costs of some counterterrorism (Donohue 2008), we must consider homegrown in critical depth. Identity framings in this context can, however unintentional, reinforce boundaries of assumed foreignness by asserting exclusionary us/them representations. This generalises individuals along categories that act as a source of insecurity for those who fall into such groupings but have no relationship to terrorism. Associations of danger with characteristics that have no intrinsic relationship to violence, such as race, religion, and immigration, unacceptably marginalise innocent actors from a secure sense of belonging.

To trouble 'homegrown' as a signifier requires that we draw on critical sensibilities, encouraging us to not take the present as given, the past as known, or the future as predetermined.[46] By viewing labels as always open to *re*construction, being critically reflective opens ways to counter terror beyond 'standard policing and military responses' (Piazza 2009: 77). This does not ignore problem solving, but encourages us to problem solve by destabilising unnecessary limitations of identity. The increased alienation of, and violence against, minority groups underscores the urgency to disrupt temporal and geographic identity signposts. It is hoped that this will mitigate generalisations of danger that blur rather than clarify, focusing instead on strategies that are inclusive of long term security and social justice.

Counterterrorism discourse and practice is connected to broader social and political relations, and as explained by Floris Vermeulen, local counterterrorism practice 'quickly devolves into a complicated, multiplex

---

[46]   See Agathangelou and Ling, (2009); Barkawi and Laffey (2006); Chowdhry and Nair (2006); Hindess (2007); Jackson *et al.* (2011); Mignolo (2010); Persaud (2006); and Shapiro (1997).

discussion about immigration, belonging, citizenship, Islam, and the position of Muslim communities in Western cities' (2014: 304). In this context representations of threat are used in legitimation struggles around exceptional policy by positioning suspects as somehow unlike 'us', as less human (Woods *et al.* 2013). The danger may be viewed as *home*grown, but the processes of boundary drawing that position homegrown as an entity (Abbott 1995; Albert *et al.* 2001) do so by distancing actors within. Distancing creates 'conditional selves' and exacerbates the alienation and insecurity of those with no relation to terror.[47] In this sense representations of threat counterproductively contribute to the idea that Islam is under attack by the West in how 'Fear and distrust, especially against innocent Muslim Americans, can easily be sown and lead to a cycle of oppression that serves to validate jihadist claims' (Rosler 2010: 66). This narrative built on complex intersections of time, space, and identity connects local conflict with global discourses, for example how in Chechnya militants are both 'grounded in a post-colonial conflict with Russia' as well as struggles in a 'global war between Islam and the West' (Swift).[48]

A key point to consider is how social and political tensions are aggravated *even when* stated policy aims are to avoid such consequences. This can be seen in an attention to 'self-radicalisation within the United States within immigrant communities' (Lister and Cruickshank: 2013) whereby the focus becomes 'immigrants, born in Western countries, who become radicalised' (King and Taylor 2011: 604). Such discourses damagingly merge 'immigrant'[49] with 'terrorist',[50] excluding from the homegrown category violence that is perpetrated by actors such as Dylan Roof. While Roof used terrorising violence in a political way, the perpetrator, target, and victims do not 'fit' simplistic categories of mainstream othering based on a racialisation and foreignisation of threat.

When the meaning of 'homegrown' is not self-evident, we must ask at what point does 'home' become associated with *non*-belonging, and 'grown' become associated with violent radicalisation?[51] How long is *long enough* to be considered part of the referent in need of protection rather than part of an other under suspicion? When the 'homegrown' other is stabilised by boundaries around migrant, immigrant, and/or Muslim categories, actors with no relation to violence but who identify with such groups are positioned as a

---

[47]   One article cites 'Alienated Muslim youths are considered by scholars and policymakers alike to be the primary source of homegrown terrorism.' (Horn 2007).

[48]   On this see Hunter and Heinke (2011), Sageman (2008:148).

[49]   This is not to negate positive references to immigrants (Mantri 2011).

[50]   See Huysmans and Buonfino (2008) on immigration in this context.

[51]   On radicalisation and grievance see McCauley and Moskalenko (2008) (radicalisation is a contested concept).

threat. Such boundary drawing is counterproductive and unethical, not least given the creation durable inequalities (Tilly 1998). Different contexts inevitably demand different considerations. At the same time, even cursory observations into how conditional (non)belonging plays out in different settings can provide important examples, as in an account of externalisation by Kurdish-German journalist Mely Kiyak:

> KIYAK: In Germany, we always talk about a special group of people. Although they are Germans, we still say they are - in Germany, we call it Auslaender, the foreigner. Although these people have a German passport, they are still the Auslaender. We do not really use this term when we mean Italian people or Spanish people, but we specially use this term for people from - coming from Turkey or the Arabic countries. They cannot really reach the status of being a normal German just because of this term.

> CORNISH: How long do you have to be in Germany before you're not considered an Auslaender?

> KIYAK: Until you die [laughter] (Cornish 2015).

This foreignising of the other along temporal lines in the German context may resonate with experiences in other areas, the significance of which would be even more problematic with respect to counterterrorism. As observed in studies on Irish and Muslim suspect communities in the UK (Nickels *et al.*, 2012: 351; Lynch 2013), the terrorist threat is often obscurely positioned, linked to ideas of 'inside' while simultaneously depending on misperceptions of the foreigner. Recent US policy focus also implies that sources of terror are based in places beyond the self, 'from North Africa to South Asia' (Obama 2013),[52] with one spatialisation of threat regarding terrorist detection stating 'Industrialised countries face the challenge of spotting international terrorists at points of entry and homegrown terrorists *on their borders*' [emphasis added] (Koc-Menard 2009). An implication from such constructions may be that the targeting of threats continues to be based on assumptions of spatial belonging and movement, even as such belonging and movement is far from predictable. Processes of boundary construction position certain actors as conditional selves that are somehow forever on the perimeter of full belonging, with an analysis of 'homegrown' perhaps providing a useful empirical snapshot to consider.

---

[52]   On spatial imaginaries see Fisher (2014).

Analysing 'Homegrown'

The homegrown other has been observed as dangerous and unexceptional, with imminent plots coming from 'unremarkable' (Silber and Bhatt 2007: 5) individuals with 'normal American lifestyles' (Pregulman and Burke 2012: 3). To get a better sense of how homegrown has been constructed, a preliminary analysis of academic and policy-focused discourse presents observations including the difficulty of defining homegrown, the separation of homegrown from other domestic and international terrorists, the presentation of homegrown as largely synonymous with jihadi extremism, and the focus on Muslim communities – all of which contribute to an ambiguous yet consequential representation of homegrown as a 'conditional self'.

Definitional dilemma

Possible consequences of insecurity from constructions of 'conditional selfhood' may be exacerbated by the confusion of defining terrorism, with boundaries around homegrown, domestic, and international labels increasingly blurred.[53] As stated by Hinkkainen Kaisa, 'there is an increasing amount of literature about homegrown terrorism, but often without a clear definition of what this constitutes' (2013:163). *Homegrown terrorism* has been defined as 'extremist violence perpetrated by U.S. citizens or legal U.S. residents, and linked to or inspired by Al Qaeda's brand of radical Sunni Islamism' (Pregulman and Burke 2012, 1); *homegrown extremists* as 'radicalized groups and individuals that are not regularly affiliated with, but draw clear inspiration and occasional guidance from, Al Qaeda core or affiliated movements' (Ibid.); and *homegrown extremism* as 'terrorist activity perpetrated by U.S. legal residents and citizens' (Nelson 2010). Official websites have presented 'homegrown violent extremism (HVE)' ('Countering...' 2015), research has identified 'self-radicalized, homegrown criminal extremists' (Carter and Carter 2012:146), and a US Senate report has referred to 'Homegrown Islamist Radicalization' ('Majority and Minority...' 2012).[54]

The challenge of multiple labels and definitions is not exclusive to the US, for example in the UK context an individual is *U.K.-based* 'only if he or she is a British citizen living in the United Kingdom or was a long-term resident of the United Kingdom during that period, regardless of immigration status', with *U.K.-related* meaning 'if he or she is a British citizen living outside the United Kingdom, or a foreign citizen who visited the United Kingdom or participated

---

53   On conceptualising terrorism see Schmid (2004).
54   See also the DHS and FBI, Joint Intelligence Bulletin, 'Use of Small Arms: Examining Lone Shooters and Small-Unit Tactics,' 16 August 2011, 3.

from abroad in a plot targeting the United Kingdom' (Barbieri and Klausen, 2012:414, note 17). Efforts have been made to specify 'homegrown', but it has also been acknowledged that any assemblage of specifics 'would constitute such a strict criterion that hardly any organization would fit the category' (Bjelopera 2013: 163). Indeed, a key observation is that 'Homegrown violent jihadist activity since 9/11 *defies easy categorization*' and 'No workable general profile of domestic violent jihadists exists' (Ibid., 2). If there is a common definitional conclusion it seems to be that the homegrown threat 'is very difficult to define' (Giuliano 2011).

An arguably productive complication of defining homegrown along varied iterations are recent articles on right wing extremism (Shane 2015),[55] with another report on 'jihadist terrorists' that references homegrown explaining just '0.007 percent of Muslims in the United States have been involved in domestic terror plots since 9/11' (Gilson 2013). Considering that 'little research has so far examined the alleged distinctiveness of homegrown terrorism empirically' (Kaisa 2013: 157), homegrown actors are said to have limited financial support and technical know-how (Bjelopera 2013: 6; Mueller and Stewart 2012:109), and that the number of 'indicted extremists' has gone from 33 in 2010 to nine in 2013 (Bergen 2013), definitional inconsistencies add further confusion to an already difficult threat assessment. With terrorism positioned in modern governmentalities as a 'risk beyond a risk' (Aradau and Van Munster 2007, 102), the consequential ambiguity of threat identification is exacerbated, not clarified, by identifiers such as domestic, international, and homegrown.

Domestic, international, homegrown

The argument that 'maintaining an artificial separation between domestic and international terrorist events impedes full understanding of terrorism and ultimately weakens counterterrorism efforts' is a compelling one (LaFree *et al.*, 2006: 6). At the time of research the FBI referenced 'domestic terrorist threats' as eco-terrorists/animal rights extremists, lone offenders, sovereign citizen movement, and anarchist extremism, separating homegrown Islamist extremist threats from 'domestic'.[56] Examples of domestic terrorism included the 'Oklahoma City bombing' and Eric Rudolph, but not 'Jihad Jane' Colleen La Rose or Nidal Hasan. Following the April 2013 Boston bombings, White House press releases stated that earlier investigations indicated no clear line if activity was 'foreign or domestic' (Carney 2013; Brennan 2013). One article was titled 'Boston Bomb Suspect: It Was Just Us' (Staff Writer 2013), but

---

[55]   See also New America Foundation, 'Deadly Attacks Since 9/11', accessed 22 July 2015 via http://securitydata.newamerica.net/extremists/deadly-attacks.html.

[56]   See FBI website 'Domestic terrorism,' http://www.fbi.gov/wanted/dt/.

assumptions of terrorists as foreign (Bulley 2008)[57] remained even as Tamerlan and Dzhokhar Tsarnaev had spent their adult lives in the US, and even as Chechen leader Ramzan Kadyrov said 'We don't know the Tsarnaevs, they did not live in Chechnya' (Locker *et al.* 2013). Responses to the attack were not focused on Americans, but on how 'Chechens are now going to be seen as bad people' (Golovnina 2013), alluding to how a historical demonisation of the 'Chechen other' can persist over time and across space (Russell 2005; Swift 2010).

Existing research has helpfully engaged with this difficulty of defining types of terrorism (e.g. Crone and Harrow 2011; Kleinmann 2012), and more research needs to be done on both disrupting threat labels and examining non-jihadist threats: 'Terrorist threats against U.S. officials and police that have nothing to do with Islamist militancy are surely also worthy of the scrutiny of Congress, but neither the Senate nor House homeland security committee, nor it seems any other congressional committee, has examined the issue in any detail since 9/11' (Bergen and Lebovich 2011). The point is not that there are no sources of insecurity self-identifying (mistakenly) with Islam. Rather, that to achieve greater security we must not let 'homegrown' enable a silencing of non-jihadist sources of terror, nor must we allow a silencing of the insecurities that stem from how counterterrorism affects those with no relation to violence.

Homegrown as 'jihadist'

Even as threats from 'homegrown violent extremists' increasingly reference both jihadist and non-jihadist extremism, there has been a particular attention to, and ongoing consequence of, constructions of terrorism as Islam-related (Jackson 2007; Croft 2012). These constructions of terrorism limit our possibilities for effective security because they encourage an almost primordial essentialisation of the terrorist, restricting our ability to conceptualise terrorism as a method of violence employed by state *and* non-state actors, by far right *and* extremist Islamist contingents. Some research has underscored the threat of far right wing militants (Perliger 2013; Taylor *et al.* 2013), but in broader public arenas such research has been received as 'outrageous' (Scarborough 2013), and 'fears [have] focused on Muslim "homegrown" terrorism"' (Brooks 2012).[58] There are also competitions over resources, with questions 'what about the homegrown extremists?' in discussions of US State Department 'foreign "Community Engagement and

---

[57]   For example, a 'Saudi man…running from the scene' was under scrutiny in part due to 'his nationality' (Eligon and Cooper), and someone was 'allowed to reenter the U.S. despite having a lapsed visa' (Carney 2013a).

[58]   On radicalisation focused on homegrown as jihadist-related see Gartenstein-Ross and Grossman (2009).

Resilience" projects' (Mirahmadi 2014), and competing understandings 'over what is international and what is domestic terrorism' (Kaisa, 2013, 159).[59] Statements that 'there is no real strategy to counter the homegrown threat' (Temple-Raston 2010) continue to be convincing, and while efforts to coordinate local and federal law enforcement may be well intentioned, receiving Suspicious Activity Reporting (SAR) training and the 'If You See Something, Say Something™' campaign ('Preventing...' 2015) do not on their own indicate effective security practice.

Instead, such practices may be counterproductive in how they can enflame insecurity from racial profiling, hate crimes, xenophobia, and Islamaphobia. Discourses that securitise Islam (Croft 2012) reinforce assumptions of terrorism as Jihadi-related, with those identified as Muslim mistakenly seen as somehow more innately terroristic than others. Immigration in this context amplifies such assumptions as the degree of threat is assessed in terms of temporal and spatial connections to 'foreign' places. '[N]ative-born European(s)' are positioned alongside representations of 'new generations of Western Muslims,' (Barbieri and Klausen 2012: 418) with the 'quintessential 'homegrown network' composed of individuals born in the West...who embraced a radical Homegrown Jihadist Terrorism in the U.S.' (Vidino 2007:2). Engaging in 'homegrown jihadism' may relate to Al Qaeda's 'long-standing subversion of migrant Muslims in the West' (Acharya and Marwah 2008), but such subversion is not *all* migrants, and the temporality of when and how someone identifies with a particular collective is not an indicator of threat. Discourses that link nonviolent communities to threats can in fact be highly destabilising, as seen in the late August 2015 attacks against migrants in Germany (Chambers 2015). Constructions of homegrown as Islamic and immigrant-related contribute to everyday insecurity for those connected to such categories but with no relation to violence, exacerbate a narrative of the West being at war with Islam, and increase the threat from anti-government extremists in how 'Immigration further fuels nativist instincts and hostility toward a federal government' (Jenkins 2012: 9).

Critical attention to constructions of homegrown as overwhelmingly 'jihadist' is not to disregard insecurity from actors espousing extremist jihadist views. Rather this is to encourage a focus on counterproductive consequences from exclusionary threat construction to better understand how phrases such as 'Muslim community' may position Islam as from 'outside' and marginalise innocent individuals as suspect, even if the intention was to *help*.

---

[59]   As a counter example they state 'Americans have been involved in terrorist activities' (Kaisa, 2013, p. 159).

'Muslim communities'[60]

> When I testified before the current House Homeland Security subcommittee on Counter-Terrorism and Intelligence chaired by Rep. King in March 2010 on 'Working with Communities to Disrupt Terror Plots', I specifically warned that solutions like Rep. King's counter-productively 'securitise the relationship' between communities and law enforcement by presenting communities with only two avenues, either as suspects or sources to report on suspects (Elibiary 2013).

It is not difficult to interpret a majority of representations of homegrown as being built upon misplaced assumptions of homegrown terrorism as almost exclusively from Muslim communities. Even as some attention to community may be well-intentioned, related discourse and practice such as 'community based policing' ('Preventing Terrorism and Countering...', 2014) can marginalise actors by positioning certain groups as needing more protection *and* more surveillance, as observed in the destructive New York Police Department surveillance programme ('End of NYPD Muslim Surveillance...' 2014). This mistakenly establishes communities as 'suspect' rather than part of the self, and is an example of how identity framings can negatively position perceived difference as danger (Hillyard 1993; McGovern and Tobin 2010; Heath-Kelly 2012; Hickman *et al.* 2012; Eriksen 2012; Breen-Smyth 2014).

By attending to the creation of suspect communities we can see how linkages between homegrown and immigrant intensify sentiments of alienation by connecting counterterrorism with the very narratives that counterterrorism is supposed to counter: that the West is at war with Islam. Insecurity is increased here in two ways: one, by the narrative being 'persuasive enough to motivate a small but disturbing number of American citizens and legal residents to take up arms to prevent further perceived assaults on Muslims' (Pregulman and Burke 2012, 4), and two, in how attacks 'by homegrown groups' heighten 'distrust toward the Muslim population among ethnic Europeans and, consequently, Muslims' sense of exclusion from mainstream society' (Vidino 2007, 589).

In looking at recent US discourse, some redirection may be under way, but more research is needed. The 2015 U.S. National Security Strategy (NSS) does not mention Muslim once, and Islam only twice, compared to the 2010

---

[60] It is not my intention to speak for anyone self-identifying with the Muslim community, and it must be noted that listening to individuals directly affected by intersections of threat construction and ideas of Muslim community should arguably be the core focus of future analysis.

NSS that focused on the need to 'build positive partnerships with Muslim communities around the world' to 'protect our homeland' (Brennan 2010). While this represents an effort to form a united front against terrorism, on the other hand such discourse also points to an interesting dynamic linking 'homeland' with communities seen as external to the self. Earlier discourse positioning the best defence against 'recruitment to jihadist terrorism in the Muslim-American community' as being 'the Muslim-American community' (Jenkins 2010: 3) and stating a need for 'positive partnerships with Muslim communities' (Obama 2010) is similar to calls for law enforcement to work 'with the Muslim-American community to identify signs of radicalisation' ('Background Briefing...' 2013): that 'the best way to prevent violent extremism inspired by violent jihadists is to work with the Muslim American community' (Obama 2013). It has been said that the US should learn from the British by focusing on 'law enforcement officers who come from a community that may be vulnerable to terrorist penetration' (Dryer 2007), but such 'community-based' efforts remain under-scrutinised and highly controversial.

As seen in mixed reactions to British Prime Minister David Cameron's 2015 speech on 'extremist ideology' (Gani 2015), in addition to important critiques on the exclusion of Muslim voices more broadly (Shafiq 2015), it is essential to consider the ways that such policies are counterproductive for security. Efforts at inclusivity are made, asserting that 'Muslims are a fundamental part of the American family' (Obama 2013). But 'community' can also signal exclusionary difference and conditional belonging. Even as Obama states that 'violent extremism is not unique to any one faith', he also says that 'our best partners in protecting vulnerable people from succumbing to violent extremist ideologies are the communities themselves', so that Muslim communities can 'protect their loved ones from becoming radicalised' (Obama 2015).[61] A continued focus on Muslim communities targets these actors with suspicion (they are externalised from full belonging with the self) *and* with responsibility (they are responsible for protecting the self). In this sense reference to Muslim communities contributes to an ongoing marginalisation that can simultaneously be seen as a cause and consequence of the insecurities identified by countering violent extremism discourse. As stated by Bjelopera (2013, 4), 'the prevention of terrorist attacks would require the cooperation and assistance of American Muslim, Arab, and Sikh communities' but 'Muslim, Arab, and Sikh Americans recognised the need to define themselves as distinctly *American* communities' [emphasis added].

Insecurity is evident in that 'Muslim community activists fear that law enforcement coerces immigrants into becoming informants' (Bjelopera 2013,

---

[61]   With thanks to an anonymous reviewer for recommending this speech.

5) and in how terrorist events are 'likely to exacerbate many of the difficulties that Muslim Americans have faced since 9/11...official discrimination by government agencies, violent hate crimes against persons and property, blatantly prejudicial legislative efforts targeted at Muslim religious practices and subtle societal discrimination that impacts employment, housing, and other attributes' (Belt 2013). Further complicating insecurity is how advocates of violence *also* refer to the Muslim community in the 'ideological desire to protect the Muslim community, which they believe is under attack by the West' (Pregulman and Burke 2012, 2). Opposing voices represent competing notions of 'Muslim community', and what it means to belong.[62]

Assumptions of difference as danger damage individuals unrelated to violence even as both 'scholars and law enforcement officials have noted that no workable general profile of domestic violent jihadists exists' and that 'generalising about the individuals involved is problematic' (Bjelopera 2013, 2, 25):

> From Qur'an burning in Florida to legislation banning the veil in France, a growing number of American and Europeans view Islam as subversive value system. This fear informs research as well, with a recent report from one Washington think-tank describing Islam as 'threat masquerading as a religion' and warning against 'U.S. leadership failures in the face of Shari'ah'. Some of this paranoia reflects nativist impulses, to be sure (Swift n.d.).

We are 'repeatedly warned about the growing alienation of American Muslims' (Vidino 2007:12) but threat labels continue to be established through oblique configurations of conditional belonging, marginalising certain individuals from full inclusion. This marginalisation is observed in the use of identity qualifiers that in one context may be celebrated as 'diversity' but in the realm of counterterrorism mark a dangerous, foreign other: a *Pakistani American* named Farooque Ahmed, an *American of Nicaraguan descent* named Antonio Martinez, and '*Pennsylvania-based* Emerson Begholly' (Giuliano 2011). Mohammad Abdulazeez, responsible for the Tennessee 2015 shootings, is described as 'Kuwait-born...a naturalised US citizen who lived most of his life in Chattanooga', instead of an 'American from Chattanooga' (Bertrand 2015).[63] Individuals are distanced from the self by focusing on qualifiers aligned with us/them discourses linked to discriminatory images of the Global South, 'a territory peopled by "Others" that can be labelled as "uncivilised",

---

[62] See Soderman (2014).
[63] Adding to the ambiguity of identification was how the FBI referred to it as 'domestic terrorism' and 'homegrown violent extremist' (Bacon, 2015).

"traditional", "irrational", and "violent", much as they were two centuries ago' (Göl 2010:2). Discursive efforts to distance violence at home from the home can thus reinforce damaging and orientalist tropes of temporal and civilisational underdevelopment, the resultant insecurities of which affect those with no relation to violence.

Reconsidering 'Homegrown'

> Homegrown terrorism has not generated much comparative literature and it seems that the debate of old and new terrorism has fused the concepts of international Islamist terrorism and homegrown terrorism into one. I, however, argue that homegrown terrorism is more similar to the other types of domestic terrorisms in Western Europe rather than its international Salafi counterpart. (Kaisa 2013, 165)

Given public imaginaries of 'foreign fighters' streaming to Syria and alarmist political discourse around the Mediterranean refugee crisis, communities that are already externalised from full belonging with the self in need of protection are increasingly seen as a target audience for counterterrorism. In the context of homegrown threats such targeting may prove problematic in a number of ways, and 'mischaracterising and inflating the Muslim homegrown American threat could prove self-defeating to the country's efforts to defend against it' (2011, 45). As argued by Elibiary, 'after facilitating more than 100 events of cooperation across our country between Muslim community members and the FBI in homegrown terrorism investigations, it is clear to me today that radicalisation is an individual or small group phenomenon that sometimes requires a community-based solution but *is never a community-level problem*' (2013) [emphasis added].

Despite some well-intentioned community-based efforts, in addition to an awareness of how counterterrorism can be counterproductive, authorities are still 'unable to conceive any coherent policy that would preemptively tackle the issue of radicalisation, preventing young American Muslims from embracing extremist ideas in the first place' (Vidino 2007, 14). Some techniques even continue to exacerbate *in*security (Vidino 2007; Bjelopera 2013), the most recent iteration being a 29 October 2015 story on how an undercover NYPD officer converted to Islam in order to spy on Muslim students. The acute insecurity experienced by one individual is powerfully explained through what could be seen as an example of damaging, conditional belonging: 'I grew up here. To have this happen because of your religion, or your political views, it's scary. You feel alienated. And you don't feel like this is your home' (Stahl 2015).[64]

The above is indicative of how security practice produces experiences of insecurity, and while 'Some may recoil at grouping right-wing, single-issue, and left-wing terrorists with militant jihadists...there are several benefits to promoting a more comprehensive assessment of the domestic terrorist threat' (Brooks 2012), such a shift may help combat counterproductive threat construction. Stereotyping rather than intelligence has been 'a key factor in the use of counter-terrorism powers' and led to 'Asian people' as 'being 11 times more likely to be stopped at UK borders' (Travis 2013), and arguments slamming 'political correctness' are an insufficient and embarrassing rationale for not stopping ineffective and counterproductive discourse and practice. The need to critically examine how we label terrorism should be felt as a matter of urgency, not least when reading how 'mothers and fathers, religious leaders and students, recent immigrants and American citizens by birth... spoke of their concerns, that their fellow Americans, and at times, their own government, may see them as a threat to American security, rather than a part of the American family' (Brennan 2010), and knowing that 'overestimating the threat could contribute to the adoption of counterproductive counterterrorism methods, especially those that threaten to alienate Muslim communities' (Brooks 2011, 9).

Given that 'we have not eliminated the sources of grievance at the United States that gave rise to Al Qaeda and could spawn other terrorist movements' (Belt), and that we have a responsibility to uphold the highest standards of social justice and human rights, it is essential to interrogate how representations of the homegrown threat in security practice may instead be influencing *in*security. This chapter is a call to consider the idea of conditional (non)belonging to support further research in this context, with the hope that this may help us find ways to combat self-defeating security discourse and practice enabled by exclusionary assumptions of (non)belonging.

The views expressed here do not represent National Defense University, the U.S. Department of Defense, or any other U.S. government entity.

References

Abbott, A. (1995) 'Things of Boundaries.' *Social Research* 62(4): 857-882.

---

[64]   See AP 'Probe into NYPD Intelligence Operations' series available from http://www.ap.org/Index/AP-In-The-News/NYPD.

Acharya, A. and Marwah, S. (2011) 'Nizam, la Tanzim (System, not Organisation): Do Organisations Matter in Terrorism Today? A Study of the November 2008 Mumbai Attacks.' *Studies in Conflict & Terrorism* vol. 34: 1–16.

Agathangelou, A. and Ling, L.H.M. (2009) *Transforming World Politics: From Empire to Multiple Worlds*. London: Routledge.

Albert, M., Jacobson, D. and Lapid, Y. (eds) (2001) *Identities, Borders, Orders: Rethinking IR Theory*. Minneapolis: University of Minnesota Press.

Aradau, C. and Van Munster, R. (2007) 'Governing Terrorism Through Risk: Taking Precautions, (un)Knowing the Future.' *European Journal of International Relations* 13(1): 89-115.

Background Briefing by Senior Administration Officials on the President's Speech on Counterterrorism Via Conference Call (2013) The White House, Office of the Press Secretary. 23 May. Available online at: http://www. whitehouse.gov/the-press-office/2013/05/23/background-briefing-senior-administration-officials-presidents-speech-co

Bacon, J. (2015) 'FBI: Chattanooga Shooter was 'Homegrown Violent Extremist'.' *USA Today*. 22 July. Available online at: http://www.usatoday. com/story/news/nation/2015/07/22/fbi-chattanooga-shooter-homegrown-violent-extremist/30513541/

Barbieri, E.T. and Klausen, J. (2012) 'Al Qaeda's London Branch: Patterns of Domestic and Transnational Network Integration.' *Studies in Conflict and Terrorism* 35(6): 411-431.

Barkawi, T. and Laffey, M. (2006) 'The Postcolonial Moment in Security Studies.' *Review of International Studies* vol. 32: 329–352.

Belt, D. (2013) 'Boston Bombing Suspects Raise New Terrorism Questions.' Interview with Anti-terrorism expert David Schanzer. *National Geographic Daily News*. 20 April. Available online at: http://news.nationalgeographic.com/ news/2013/13/130420-homegrown-terrorism-tsarnaev-brothers-boston-marathon-bombings-islam-radicalism/

Bergen, P. and Lebovich, A. (2011) 'Measuring the Homegrown Terrorist Threat to U.S. Military.' *CNN* 7. 7 December. Available online at: http://www. cnn.com/2011/12/07/opinion/bergen-terrorist-threat-military/index.html

Bergen, P. (2013) 'Hyping the Terror Threat?' *CNN*. 3 December. Available online at: http://peterbergen.com/hyping-the-terror-threat-cnn-com.

Bertrand, N. (2015) 'Marines Reportedly Sacrificed Themselves to Protect Fellow Troops in Chattanooga Shooting.' *Business Insider*. 22 July. Available online at: http://www.businessinsider.com/chattanooga-shooting-marines-sacrificed-themselves-2015-7

Biswas, S. (2004) 'The "New Cold War": Secularism, Orientalism, and Postcoloniality.' In: Chowdhry, C. and Nair, S. (eds) *Power, Postcolonialism and International Relations: Reading race, gender and class*. London: Routledge. 184-208.

Bjelopera, J.P. (2013) 'American Jihadist Terrorism: Combating a Complex Threat.' *Congressional Research Service Report for Congress*. 7-5700, R41416. 23 January. Available online at: http://www.fas.org/sgp/crs/terror/R41416.pdf

Breen-Smyth, M. (2014) 'Theorising the "Suspect community": Counterterrorism, Security Practices and the Public Imagination.' *Critical Studies on Terrorism* 7(2): 223-240.

Brennan, J. (2010) 'Securing the Homeland by Renewing American Strength, Resilience and Values.' 26 May. Available online at: http://www.whitehouse.gov/the-press-office/remarks-assistant-president-homeland-security-and-counterterrorism-john-brennan-csi

Brooks, R. (3 May 2012) 'Homegrown Terror isn't Just Islamist.' *CNN*. Available online at: http://www.cnn.com/2012/05/03/opinion/brooks-bridge-homegrown-terrorists/index.html

Brooks, R. (2011) 'Muslim "Homegrown" Terrorism in the United States: How Serious Is the Threat?' *International Security* 36(2): 7-47.

Bulley, D. (2008) '"Foreign" Terror? London Bombings, Resistance and the Failing State.' *British Journal of Politics and International Relations* 10(3): 379-394.

Carney, J. (2013) 'Press Briefing by Press Secretary Jay Carney.' 23 April. Available online at: http://www.whitehouse.gov/the-press-office/2013/04/23/press-briefing-press-secretary-jay-carney-4232013

Carney, J. (2013a) The White House Office of the Press Secretary For Immediate Release. "Press Gaggle by Press Secretary Jay Carney Aboard Air Force One En Route Mexico City, Mexico." 2 May. Available online at: http://www.whitehouse.gov/the-press-office/2013/05/02/press-gaggle-press-secretary-jay-carney-aboard-air-force-one-en-route-me.

Carter, J.G. and Carter, D.L. (2012) 'Law Enforcement Intelligence: Implications for Self-Radicalized Terrorism.' *Police Practice and Research* 13(2): 138-154.

Chambers, M. (2015) 'Germany vows to fight xenophobia after attacks on refugee home.' *Reuters.* 23 August. Available online at: http://www.reuters.com/article/2015/08/23/us-europe-migrants-germany-asylum-idUSKCN0QS0DZ20150823

Chowdhry, G. and Nair, S. (2006) 'Introduction: Power in a postcolonial world: race, gender, and class in international relations.' In: Chowdhry, G. and Nair, S. (eds) *Power, Postcolonialism and International Relations: Reading race, gender and class.* London: Routledge. 1-32.

Chuang, A. and Roemer, R.C. (2013) 'The Immigrant Muslim American at the Boundary of Insider and Outsider: Representations of Faisal Shahzad as "Homegrown" Terrorist.' *Journalism and Mass Communication Quarterly* 90(1): 89-107.

Cornish, A. (2015) 'Kurdish-German Journalist Makes Light Of Hate Mail In Spoken Word Act.' *Special Series Muslim Identity in Europe, National Public Radio.* 5 March. Available online at: http://www.npr.org/2015/03/05/391041966/german-journalist-makes-light-of-hate-mail-in-spoken-word-act

Countering Violent Extremism (2015) U.S. Department of Homeland Security. Available online at: http://www.dhs.gov/topic/countering-violent-extremism.

Croft, S. (2012) *Securitising Islam: Identity and the Search for Security.* Cambridge: Cambridge University Press.

Donohue, L.L. (2008) *The Costs of Counterterrorism: Power, Politics, and Liberty.* Cambridge: Cambridge University Press.

Dryer, D. (2007) "Terrorism in the West: Al-Qaeda's Role in "Homegrown" Terror.' Interview with Bruce Hoffman. *Brown Journal of World Affairs.* Washington DC. 22 February. 13(2).

Elibiary, M. (2013) 'Boston bombings and the radicalised homegrown terrorist.' *The Washington Post.* 30 April. Available online at: http:// extremisproject.org/2013/05/news-from-north-america-wednesday-1st-may-2013/

Eligon, J. and Cooper, M. (2013) 'Blasts at Boston Marathon Kill 3 and Injure 100.' *The New York Times.* 15 April. Available online at: http://www.nytimes. com/2013/04/16/us/explosions-reported-at-site-of-boston-marathon. html?pagewanted%253Dall&_r=0

'End of NYPD Muslim Surveillance Program Applauded' (2014) *CBS News New York.* 16 April. Available online at: http://newyork.cbslocal. com/2014/04/16/end-of-nypd-muslim-surveillance-program-applauded

Eriksen, E. (2012) 'Why the UK PREVENT Strategy Does Not Prevent Terrorism.' *E-International Relations.* 5 September. Available online at: http:// www.e-ir.info/2012/09/25/why-the-uk-prevent-strategy-does-not-prevent-terrorism/

Finighan, A. (2014] 'UK bill: Passport to statelessness?' *Al Jazeera News.* 13 May. Available online at: https://uk.news.yahoo.com/uk-bill-passport-statelessness-204930164.html#4lvwWEd

Fisher, K.M. (2014) 'Spatial and temporal imaginaries in the securitisation of terrorism.' In: Jarvis, L. and Lister, M. (eds) *Critical Perspectives on Counter Terrorism*, Basingstoke: Routledge. 56-76.

Fisher, K.M. (2015) *Security, Identity, and British Counterterrorism Policy.* Basingstoke: Palgrave Macmillan.

Gani, A. (2015) 'Cameron's extremism speech gets mixed response from Birmingham Muslims.' *The Guardian.* 20 July. Available online at: http://www. theguardian.com/politics/2015/jul/20/camerons-extremism-speech-gets-mixed-response-from-birmingham-muslims

Gartenstein-Ross, D. and Grossman, L. (2009) 'Homegrown terrorists in the U.S. and U.K.: An Empirical Examination of the Radicalisation Process.' *FDD Press Foundation for Defense of Democracies.* Available online at: http://www.defenddemocracy.org/content/uploads/documents/HomegrownTerrorists_USandUK.pdf

Gilson, D. (24 April 2013) 'Charts: How Much Danger Do We Face from Homegrown Jihadist Terrorists?' *Mother Jones.* Available online at: http://www.motherjones.com/politics/2013/04/charts-domestic-terrorism-jihadist-boston-tsarnaev

Giuliano, M.F. (2011) FBI speech by Mark F. Giuliano, Assistant Director, Counterterrorism Division, Federal Bureau of Investigation. The Washington Institute for Near East Policy Stein Program on Counterterrorism and Intelligence. Washington DC. 14 April. Available online at: http://www.fbi.gov/news/speeches/the-post-9-11-fbi-the-bureaus-response-to-evolving-threats

Göl, A. (2010) 'Editor's introduction: views from the "Others" of the War on Terror.' *Critical Studies on Terrorism* 3(1): 1-5.

Golovnina, M. (2013) 'Tsarnaev homeland Chechnya: rebuilt from war, ruled by fear.' *Reuters.* 1 May. Available online at: http://www.reuters.com/article/2013/05/01/us-usa-explosions-russia-chechnya-idUSBRE94005Y20130501

Heath-Kelly, C. (2012) 'Counter-Terrorism and the Counterfactual: Producing the "Radicalisation" Discourse and the UK PREVENT Strategy.' *British Journal of Politics and International Relations* 15(3): 394-415.

Hickman, M., Thomas, L. Nickels, H. and Silvestri, S. (2012) 'Social cohesion and the notion of "suspect communities": a study of the experiences and impacts of being "suspect" for Irish communities and Muslim communities in Britain.' *Critical Studies on Terrorism* 5(1): 89–106

Hindess, B. (2007) 'The Past is Another Culture.' *International Political Sociology* 1(4):325–338.

Horn, H. (2007) 'Neglect's Costs: Turkish Integration in Germany.' *Harvard International Review.* Available online at: http://hir.harvard.edu/archives/1602

Horwitz, S. (2013) 'FBI's Comey focused on violent home-grown extremists, global spread of terrorism.' *The Washington Post.* 19 September. Available online at: http://www.washingtonpost.com/world/national-security/fbis-comey-focused-on-violent-home-grown-extremists-global-spread-of-terrorism/2013/09/19/dda2cfd2-2159-11e3-966c-9c4293c47ebe_story.html

Heinke, D. H. and Hunter, R. (2011) 'Radicalisation of Islamist Terrorists in the Western World.' *FBI Law Enforcement Bulletin* 80.9: 25-31. Available online at: http://works.bepress.com/daniel_heinke/38

Huysmans, J. and Buonfino, A. (2008) 'Politics of Exception and Unease: Immigration, Asylum and Terrorism in Parliamentary Debates in the UK.' *Political Studies* 56(4): 766-788.

Jackson, R. (2007) 'Constructing Enemies: "Islamic Terrorism" in Political and Academic Discourse.' *Government and Opposition* 42(3): 394–426.

Jackson, R., Jarvis, J., Gunning, J. and Breen-Smyth, M. (2011) *Terrorism: A Critical Introduction*. Basingstoke: Palgrave Macmillan.

Jenkins, B.M. (2010) 'Testimony No Path to Glory: Deterring Homegrown Terrorism.' CT-348 Testimony presented before the House Homeland Security Committee, Subcommittee on Intelligence, Information Sharing and Terrorism Risk Assessment. 26 May. Available online at: http://www.rand.org/pubs/testimonies/CT348.html

Jenkins, B.M. (2013) 'New Challenges to U.S. Counterterrorism Efforts: An Assessment of the Current Terrorist Threat.' CT-377 Testimony to Senate Homeland Security and Governmental Affairs Committee. 11 July. Available online at: http://www.rand.org/pubs/testimonies/CT377.html

Kaisa, H. (2013) 'Homegrown Terrorism: The Known Unknown.' *Peace Economics, Peace Science and Public Policy* 19(2): 157-182.

King, M. and Taylor, D.M. (2011) 'The Radicalisation of Homegrown Jihadists: A Review of Theoretical Models and Social Psychological Evidence, Terrorism and Political Violence.' *Terrorism and Political Violence* 23(4): 602-622.

Koc-Menard, S. (2009) 'Trends in Terrorist Detection Systems.' *Journal of Homeland Security and Emergency Management* 6(1): 1-13.

LaFree, G., Dugan, L., Fogg, H.V. and Scott, J. (2006) 'Building a Global Terrorism Database.' University of Maryland. 27 April. Available online at: https://www.ncjrs.gov/pdffiles1/nij/grants/214260.pdf

Lister, T. and Cruickshank, P. (2013) 'Dead Boston bomb suspect posted video of jihadist.' *CNN*. 22 April. Available online at: http://www.cnn.com/2013/04/20/us/brother-religious-language/index.html

Locker, R., McCoy, K. and Zoroya, K. (2013) 'The Chechen People: A Fierce Resistance.' *USA Today*. 20 April. Available online at: http://www.usatoday.com/story/news/world/2013/04/19/russia-chechnya-terror-caucasus/2095995/

Lynch, O. (2013) 'British Muslim youth: radicalisation, terrorism and the construction of the "other."' *Critical Studies on Terrorism* 6(2): 241-261.

Majority and Minority Staff Senate Committee on Homeland Security and Governmental Affairs (2012) 'Zachary Chesser: A Case Study in Online Islamist Radicalisation and Its Meaning for the Threat of Homegrown Terrorism.' Available online at: http://www.hsgac.senate.gov/imo/media/doc/CHESSER%20FINAL%20REPORT(1).pdf

Mantri, G. (2011) 'Homegrown Terrorism: Is There an Islamic Wave?' *Harvard International Review* 33(1): 88-94.

McCauley, C. and Moskalenko, S. (2008) 'Mechanisms of Political Radicalisation: Pathways Toward Terrorism.' *Terrorism and Political Violence* 20(3): 415-433.

McGovern, M. and Tobin, A. (2010) *Countering Terror or Counter-Productive? Comparing Irish and British Muslim Experiences of Counter-insurgency Law and Policy Report*. Lancashire: Edge Hill University.

Mignolo, W. (2010) *The Darker Side of the Renaissance: Literacy, territoriality, and colonisation*. Ann Arbor: University of Michigan Press.

Mirahmadi, H. (2014) 'Addressing the Homegrown Terrorism Threat.' 10 January. Available online at: http://www.rollcall.com/news/addressing_the_homegrown_terrorism_threat_commentary-230070-1.html

Mueller, J. and Stewart, M.G. (2012) 'The Terrorism Delusion: America's Overwrought Response to September 11.' *International Security* 37(1): 81-110.

Nelson, R. "Ozzie" (2010) 'Countering Terrorism and Radicalisation in 2010 and Beyond: A New Terrorist Threat? Assessing "Homegrown Extremism.' *Center for Strategic and International Studies.* Available online at: http://csis.org/files/publication/100121_countering_terrorism_n_radicalization.pdf

Nickels, H. C., Hickman, T,T. and Silvestri, S. (2012) 'De/constructing "suspect" communities.' *Journalism Studies* 13(3): 340-355.

Obama, B. (2010) *US National Security Strategy.* Available online at: http://www.whitehouse.gov/sites/default/files/rss_viewer/national_security_strategy.pdf [

Obama, B. (2013) President Obama's Speech. 23 May. Available online at: http://www.whitehouse.gov/photos-and-video/video/2013/05/23/president-obama-speaks-us-counterterrorism-strategy#transcript

Obama, B. (2015) Remarks by President Obama at the Leaders Summit on Countering ISIL and Violent Extremism. United Nations Headquarters New York, NY. 29 September. Available online at: https://www.whitehouse.gov/the-press-office/2015/09/29/remarks-president-obama-leaders-summit-countering-isil-and-violent.

Perliger, A. (2013) 'Challengers from the Sidelines: Understanding America's Violent Far-Right.' 15 January. Available online at: https://www.ctc.usma.edu/wp-content/uploads/2013/01/ChallengersFromtheSidelines.pdf

Persaud, R.B. (2006) 'Situating Race in International Relations: The dialectics of civilisational security in American immigration.' In: Chowdhry, G. and Nair, S. (eds) *Power, Postcolonialism and International Relations: Reading race, gender and class.* London: Routledge: 56-81.

Piazza, J. A. (2009) 'Is Islamist Terrorism More Dangerous? An Empirical Study of Group Ideology, Organisation and Goal Structure.' *Terrorism and Political Violence* 21(1): 62-88.

Pregulman, A. and Burke, E. (2012) 'Homegrown Terrorism.' *Al Qaeda and Associated Movements (AQAM) Futures Project Case Study Series,* Center for Strategic and International Studies. Case Study Number 17. Available online at: http://csis.org/files/publication/120425_Pregulman_AQAMCaseStudy7_web.pdf

Department of Homeland Security (2015) *Preventing Terrorism Results*. 13 July. Available online at: http://www.dhs.gov/topic/preventing-terrorism-results.

Organisation for Security and Cooperation in Europe (2014) *Preventing Terrorism and Countering Violent Extremism and Radicalisation that Lead to Terrorism: A Community-Policing Approach*. Available online at: http://www.osce.org/atu/111438

Rosler, A. (2010) 'Devolving Jihadism.' *Journal of Strategic Security* 3(4): 63-74.

Russell, J. (2005) 'Terrorists, Bandits, Spooks and Thieves: Russian Demonisation of the Chechens Before and Since 9/11.' *Third World Quarterly* 26(1): 101-116.

Said, E. (1979) *Orientalism*. 25th Ed. New York: Vintage Books Random House.

Sageman, M. (2008) *Leaderless Jihad: Terror Networks in the 21st Century*. University of Pennsylvania Press: Philadelphia.

Scarborough, R. (2013)'"Far Right" report outrages critics of federalism.' *The Washington Times*. 21 January. Available online at: http://www.washingtontimes.com/news/2013/jan/21/far-right-report-outrages-critics-of-federalism/#ixzz2vJ8mVj00

Schmid, A. (2004) 'Frameworks for Conceptualising Terrorism.' *Terrorism and Political Violence* 16(2): 197-221.

Shafiq, M. (2015) 'David Cameron has made it all about Muslims – without engaging us at all.' *The Guardian*. 20 July. Available online at: http://www.theguardian.com/commentisfree/2015/jul/20/david-cameron-british-muslims-extremism-speech

Shane, S. (2015) 'Homegrown Extremists Tied to Deadlier Toll Than Jihadists in U.S. Since 9/11.' *New York Times*. 24 June. Available online at: http://www.nytimes.com/2015/06/25/us/tally-of-attacks-in-us-challenges-perceptions-of-top-terror-threat.html?_r=0

Shapiro, M.J.J. (1997) *Violent Cartographies.* Minneapolis: University of Minnesota Press.

Silber, M. D. and Bhatt, A. (2007) 'Radicalisation in the West: The Homegrown Threat.' *NYPD Intelligence Division.* Available online at: http://www.nypdshield.org/public/SiteFiles/documents/NYPD_Report-Radicalization_in_the_West.pdf

Soderman, J. (2014) 'Al Qaeda calling for car bomb attacks on American soil.' 8 May. Available online at: http://www.kusi.com/story/25460447/al-qaeda-calling-for-car-bomb-attacks-on-american-soil)

Staff Writer (2013) 'Boston Bomb Suspect: "It Was Just Us."' *CNN (hosted by Fox).* 23 April., Available online at: http://fox2now.com/2013/04/23/source-boston-bomb-suspect-says-brother-was-brains-behind-attack/.

Stahl, A. (2015) 'NYPD Undercover "Converted" To Islam To Spy On Brooklyn College Students.' *The Gothamist.* 29 October. Available online at: http://gothamist.com/2015/10/29/nypd_undercover_brooklyn.php

Swift, C. (no date given) 'Western Errors in the War on Terror.' Available online at: http://christopher-swift.com/publications/western-errors-in-the-war-on-terror

Swift, C. (2010) 'Fragmentation in the North Caucasus Insurgency.' *Combating Terrorism Center (CTC) Sentinel* 3(11/12).

Taylor, M., Currie P.M. and Holbrook, D. (2013) *Extreme Right Wing Political Violence and Terrorism.* Bloomsbury Press.

Temple-Raston, D. (2010) 'Homegrown Terrorists Pose Biggest Threat.' *NPR.* 10 September. Available online at: http://www.mprnews.org/story/npr/129760267

Tilly, C. (1998) *Durable Inequality.* Berkeley: University of California Press.

Travis, A. (2013) 'Asian People 11 Times More Likely to be Stopped at UK borders, Analysis Finds EHRC Analysis Suggests Stereotyping Rather than Intelligence May be KeyFactor in Use of Counter-Terrorism Powers.' *The Guardian.* 5 December. Available online at: http://www.theguardian.com/law/2013/dec/05/asian-people-stopped-uk-borders-analysis

Vermeulen, F. (2014) 'Suspect Communities—Targeting Violent Extremism at the Local Level: Policies of Engagement in Amsterdam, Berlin, and London.' *Terrorism and Political Violence* 26(2): 286-306.

Vidino, L. (2007) 'The Hofstad Group: The New Face of Terrorist Networks in Europe.' *Studies in Conflict and Terrorism* 30(7): 579-592.

Williams, P. (2015) 'DOJ Creates New Domestic Terror Position.' *NBC News.* 14 October. Available online at: http://www.nbcnews.com/news/us-news/doj-creates-new-domestic-terror-position-n444436

Woods, C., Ross, A. and Wright, O. (2013) 'British terror suspects quietly stripped of citizenship… then killed by drones: Exclusive: Secret war on enemy within.' *The Independent.* 28 February. Available online at: http://www.independent.co.uk/news/uk/crime/british-terror-suspects-quietly-stripped-of-citizenship-then-killed-by-drones-8513858.html

11

Catastrophic Futures, Precarious Presents, and the Temporal Politics of (In) security[65]

LIAM P.D. STOCKDALE
MCMASTER UNIVERSITY, CANADA

Introduction: The 'Temporalisation' of (In)security

In Chapter III of *The Prince*, Machiavelli declares that 'it is necessary not only to pay attention to immediate crises, but to foresee those that will come and to make every effort to prevent them' (1995). While perhaps a basic political truism, this idea of actively governing the future through targeted interventions in the present has been central to the Western-led response to the terrorist attacks of 11 September 2001. Indeed, because transnational terrorism has been widely framed as a novel type of danger that is uniquely 'unpredictable in occurrence, characteristics, and effects' (Anderson 2010b: 228), the most pressing 'security issues have increasingly been defined in terms of uncertain, potentially catastrophic threats' (Aalberts and Werner 2011: 2188, 2191). In this sense, the ever-present spectre of a seemingly inevitable next attack has inscribed a radically contingent, potentially catastrophic future as the primary threat against which security measures must be oriented. This has led the very idea of 'security' to be framed in

---

[65] This paper presents a condensed version of the arguments developed in my book *Taming an Uncertain Future: Temporality, Sovereignty, and the Politics of Anticipatory Governance*, (Rowman & Littlefield, 2016). I thank the anonymous reviewer(s) from E-IR for a number of productive comments and critiques that have helped me refine the arguments developed below.

essentially temporal terms, and equated with taming this future through anticipatory action in the present. The advent of the War on Terror, in short, has 'reconfigured the politics of space into a politics of time' in the realm of (in)security governance (Kessler 2011: 2181).

To be sure, there is an anticipatory element to even the most basic understanding of security, as its pursuit necessarily involves not only responses to attacks, but the identification and ongoing prevention of future threats. The novelty of the post-9/11 reorientation of global (in)security politics along more temporal lines should thus not be overstated. Yet its practical operationalisation has had significant implications for the way political power is organised and exercised that merit critical scrutiny. Among the most notable is the extent to which the attendant institutionalisation of 'pre-emption' as a security rationality has manifested as a politics of 'exceptionalism' under which precarious political subjectivities are enacted. In what follows, I consider how this correlation between pre-emption and exceptionalism is no coincidence, as the latter can be understood as an originary function of the former that stems from pre-emption's underlying temporal imperative to tame a radically uncertain future. In other words, I argue that a pre-emptive politics of (in)security logically presupposes what amounts to a politics of exceptionalism in which the relationships between sovereign powers and the subjects they govern approximate the sort of unmediated confrontation often described in the Critical Security Studies literature informed by readings of Giorgio Agamben's work on sovereignty.

Towards this end, I begin by describing how the post-9/11 'temporalisation' of (in)security has taken the form of a politics of pre-emption in which radical uncertainty constitutes the basis for, rather than an impediment to, anticipatory action. I then consider how this requires a prioritisation of the imagination in the context of anticipatory decision-making, which in turn grants the deciding entity a significantly enhanced degree of discretionary authority. Through a brief examination of the Obama administration's drone warfare programme and the targeted killing of US citizen Anwar al-Awlaki, I subsequently consider how, in practice, this enacts precarious political subjectivities. I conclude by considering the implications of these ideas for debates about the compatibility of anticipatory security rationalities with the norms of liberal democratic governance.

Pre-emption in the Post-9/11 World

An extensive body of Critical Security Studies work has documented the recent proliferation of what can be generally characterised as 'pre-emptive' security strategies (see, for example, Amoore 2014; Stockdale 2013;

Anderson 2010a; de Goede and Randalls 2009; Massumi 2007; Elmer and Opel 2006; Cooper 2006). Addressing a wide range of issues—including, but not limited to: the indefinite detention (Ericson 2008) and targeted killing (Leander 2011) of suspected terrorists; the biometric monitoring of mobile populations (Muller 2010); the pre-emptive detention of refugees (Isin and Rygiel 2007); the anticipatory freezing of monies suspected of terrorist links (de Goede 2012); and the so-called 'Bush Doctrine' of pre-emptive inter-state war (Weber 2007)—this literature has comprehensively illustrated that a dominant trend in contemporary global (in)security governance is the turn towards a logic of pre-emption premised on intervening in the present to create alternative futures in which potential catastrophes are precluded.

Looking at this trend from a wider perspective, it can be usefully understood as a function of broader shifts in the way the present and future are framed in the contemporary global security imagination. In this respect, the present is construed as beset by a radical contingency, with the global security environment in particular characterised by the breakdown of established certainties in the face of novel threats (Kessler and Daase 2008). The consequence is that 'decision-makers are simply no longer able to guarantee predictability, security, and control' to the extent that was once believed possible (Aradau and van Munster 2008: 23). Complementary to this framing of the present, the future is understood in terms not just of radical uncertainty but of 'expected and undeniable catastrophe', typified by the proverbial next attack, with the precise moment of its emergence remaining continually unknown (*Ibid.*). In other words, the future embodies an impending catastrophe that is seemingly both inevitable—in that it *will* at some point occur—and unknowable—in that we cannot be certain *when* it will take place (Anderson 2010a: 779-80; Opitz and Tellmann 2015: 107). What more broadly results is what Paul Virilio has termed 'a culture of the imminence of disaster' (2010: 7), in which—to wax Rumsfeldian—contemporary life is haunted by the spectre of a disastrous unknown that is paradoxically known to be lurking in the future's unknowable depths.

Approaches premised upon taming these uncertainties by controlling the unfolding of the future through anticipatory interventions in the present have accordingly come to dominate the post-9/11 global politics of (in)security. In practice, this perspective implies the necessity of acting pre-emptively, even without an established base of verifiable knowledge upon which to make such decisions (Aradau and van Munster 2007: 101). In other words, radical *un*certainty about both the nature of the threat and the moment at which it might emerge is not an *impediment to* acting anticipatorily; it instead provides the very *impetus for* it, such that 'the absence of specific evidence serves as justification for action' (Elmer and Opel 2006: 481). This places governing authorities 'in the uncomfortable position of having to take drastic action in the face of an inescapably elusive, uncertain threat' (Cooper 2006: 119).

A key consequence is that established mechanisms and logics of (in)security governance nominally premised upon data collection and deliberation appear inadequate, and strategies based primarily on conjecture and speculation emerge as the most viable options (de Goede 2012). The Obama administration's use of 'predictive assessments about potential threats' as the basis for the inclusion of individuals on so-called 'no-fly lists' and other 'watchlisting' initiatives central to US counter-terrorism efforts is an instructive example here (Ackerman 2015). At a May 2015 a federal court proceeding, for instance, two officials from the US Justice Department described the watchlisting programme's chief goal as 'identifying individuals who may be a threat to civil aviation and national security' through 'predictive judgement[s] intended to prevent future acts of terrorism in an uncertain context' (quoted in *Ibid*.). This testimony provided official confirmation that decisions to place particular individuals on no-fly lists are 'based on predicting crimes rather than on records of demonstrated offenses' (*Ibid*.). And despite an FBI official telling the court that 'mere guesses or 'hunches'…are not sufficient,' a former CIA counterterrorism analyst responded by testifying that 'there is no indication that the government has assessed the scientific validity and reliability of its predictive judgments,' and as such, these judgments 'amount to little more than the 'guesses' or 'hunches' that [the FBI official] says are not sufficient' (quoted in *Ibid*.). Considered in light of the importance of watchlisting practices to US counter-terror strategy (Herman 2011), the CIA analyst's testimony and the revelations of the court proceedings more generally illustrate both how the importance of taking action in the face of radical uncertainty defines the post-9/11 politics of (in)security, and that this has a significant effect on the way decisions are made in the security context. Indeed, if the mere possibility of danger emerging at some point in the future can serve as the basis for anticipatory action in the present, the implications for the organisation and exercise of sovereign power in the security context are not insignificant.

The Future and the Imagination

To begin to understand these implications, it is useful to understand the reformulation of (in)security governance associated with a politics of pre-emption in terms of the prioritisation of the future over the present. In this respect, we have seen that a pre-emptive approach requires that some sort of anticipatory action be taken for the purpose of governing an uncertain future; yet such action must necessarily take place in the present, since this is where humans agents act. The ontology of security is thus constructed in such a way that the future itself becomes the referent to be secured—with catastrophic iterations thereof constituting the threat to be secured *against*— while the present is instrumentally constructed as the location of the interventions necessary accomplish this. In other words, the logic of pre-emption prioritises the governance of the future, which in turn implies the legitimation of potentially disruptive interventions in the present, thus effectively placing the latter in subordinate service to the former.

Importantly, this prioritisation of the future alters the epistemic basis of political action by enacting a decisional logic that relies upon 'knowledge' derived primarily from the exercise of the imagination. Indeed, as Aradau and van Munster put it, 'imagination acquires epistemic primacy in relation to the unknown' (2011: 85); and because pre-emptive interventions act upon a radically uncertain (and thus ultimately unknown) future, the imagination is vital to rendering that which is to be pre-empted intelligible and therefore actionable in the present. In other words, 'imagination is constitutive of security knowledge' (*Ibid.*, 84), and any security decision through which anticipatory action is taken will rely upon the imagination to a significant degree. The result is that a pre-emptive politics of (in)security shifts the epistemic basis for action from the realm of verifiable fact to the realm of speculation, conjecture, and suspicion (de Goede and Randalls 2009: 868). What follows is that anticipatory action is in practice based to a significant extent upon evidentiary foundations constructed by the deciding authorities (or their designated surrogates). Under a politics of pre-emption, therefore, those tasked with making anticipatory decisions are also ultimately tasked with creating the epistemic basis for these decisions through the exercise of the imagination. This has the effect of greatly enhancing the scope of discretionary authority granted to those who decide how and when to act in a pre-emptive security context.

An illustrative example can again help clarify these points. On 29 March 2011, Mohamed Hersi—a 25-year-old Somali-Canadian—was arrested by the Royal Canadian Mounted Police (RCMP) as he tried to board a flight from Toronto to Cairo via London. Hersi was charged, tried, and ultimately became the first Canadian convicted of 'attempting to participate in a terrorist activity'—an offence for which he was sentenced to the maximum of 10 years in prison. Like most Western states in the post-9/11 era, Canada has adopted a pre-emptive approach to governing terrorism premised explicitly upon 'protecting Canadians from terrorist acts before they occur' (Government of Canada 2011: 32). The country's anti-terrorism guidelines have thus been developed to allow security agencies to act in such a way that threats are 'dealt with on more of an *a priori* basis rather than more of a post facto basis' (Svendsen 2010: 320). The RCMP was operating under this regime when Hersi was detained, and the inchoate nature of his offence also highlights this pre-emptive ethos.

The RCMP alleged that Hersi intended to proceed from Cairo to Somalia to train as a militant with Al-Shabaab, a group designated as a terrorist organisation under Canadian law. At the time of the arrest, however, the RCMP offered a rather candid admission that their investigation 'did not indicate that the suspect was a direct threat to his country or Canadians' at the time of his detention (Teotonio 2011). In other words, the threat that was pre-empted by Hersi's arrest did not tangibly exist at the moment he was detained. He was, rather, subjected to an intervention by the Canadian state based on an *imagined future threat* that he *may have one day* posed. This conjectural aspect of the case was used unsuccessfully in Hersi's defence at trial, after which his lawyer promised to appeal, claiming that the conviction relied on speculative evidence that was 'so far removed from reality as to make it a thought crime' (Humphreys 2014).

Of course, the decision to detain Hersi was presumably not entirely arbitrary; and liberal legal norms have long included conspiracy provisions permitting the anticipatory arrest of those taking preparatory steps to commit a serious crime who have not yet followed through. So while it is certainly arguable whether the Hersi case falls outside the normative boundaries of a liberal legal order, it still usefully illustrates how the praxis of pre-emptive security operates through the privileging of imagination as the epistemic foundation for action, which in turn enhances the discretionary authority of those tasked with deciding upon such action by requiring them to construct that foundation. When considered in a broader context, this suggests that pre-emptive security measures function through a paradigm of political power in many ways reminiscent of that which is associated with a politics of 'exceptionalism.'

Pre-emption and Exceptionalism

The Critical Security Studies literature includes myriad engagements with the work of Carl Schmitt, Giorgio Agamben, and others who have extensively theorised the concept of political 'exceptionalism'. Described most simply, exceptionalism denotes a condition where the prevailing legal order is effectively annulled and a more arbitrary, unconstrained form of power vested in a particular person or office emerges (Schmitt 2005: 12; Huysmans 2004). As I have argued elsewhere (Stockdale 2013), political exceptionalism can therefore be understood as characterised by two core components: the suspension of the juridical order (Agamben 2005: 23), and a 'decisionist' paradigm of political authority (Schmitt 2005: 33). With respect to the first, a politics of exceptionalism involves freeing the highest form of legitimate political authority—usually the executive—from limitations imposed by the rule of law (Schmitt 2005: 11; Lazar 2006: 260). Political action thus takes a more arbitrary form, since the absence of effective legal constraints grants an enhanced degree of discretion to the acting authority. The decisionist component of exceptionalism thus follows from this, as it describes a condition where the executive does not simply apply the law, but rather is effectively 'the source of law' (Lazar 2006: 257). Under a state of exception, in other words, a decision taken by the relevant authority is purer, such that, in Schmitt's words, it 'emanates from nothingness' (Schmitt 2005: 33).

When these points are read against the discussion of pre-emptive security in the preceding sections, it becomes clear that a paradigm of political power comprising these two core elements of exceptionalism bears a close resemblance to that which accompanies the logic of pre-emption as applied to the contemporary governance of (in)security. Indeed, by prioritising the imagination and concomitantly augmenting the deciding authority's discretion, a politics of pre-emption effectively presupposes what amounts to a politics of exceptionalism by both suspending the juridical order and enacting a decisionist paradigm of political authority. It is worth considering each of these points in more detail.

Regarding the first, a corollary of pre-emption's focus on governing a radically uncertain future is that no imaginable possibility can be dismissed—which implies more specifically that no individual can be presumptively absolved of suspicion in the present (Ericson 2008). In other words, when dealing with imagined futures, there is no way to prove an accused's innocence once the imagination has been activated, since accusations relate to an act that has not yet taken place. This undermines the basis of almost any type of juridical order, since the collection and evaluation of evidence via designated channels is rendered largely impossible when dealing with imagined future events. The power to assess an individual's (future) guilt—and thus to subject her/him to (present) interventions—is therefore transferred from the mechanisms of the legal system to the whim of a designated political authority. Of course, in practice this rarely translates into the state detaining those suspected of posing a potential threat on entirely arbitrary grounds. The point, however, is that the logic of pre-emption conflicts in crucial ways with the established norms of a liberal juridical order, since the latter are not equipped to handle the extended temporal horizons and evidentiary uncertainties that inevitably accompany anticipatory action (Aradau and van Munster 2009: 697).[66] In this sense, pre-emptive approaches to (in)security

---

[66] One way this tension has been negotiated is through what some critical commentators have termed 'counter law' (Opitz and Tellmann 2015, Ericson 2008). This term refers to legislative moves aimed at embedding practices that undermine the constitutive norms of a liberal juridical order within that order. Importantly, Ericson links counter law specifically to the emergence of pre-emptive security measures, describing it as the creation of new laws and/or 'new uses of existing laws' that 'erode or eliminate traditional principles, standards, and procedures of criminal law that get in the way of

governance suspend the juridical order almost by default, since questions regarding threat, culpability, and response become the purview of sovereign political authorities.

Regarding the second core component of exceptionalism, the type of decision required by pre-emptive security strategies demands what amounts to a 'decisionist' form of political authority. In this respect, we have seen that because the futures against which anticipatory decisions are framed exist in the realm of the imagination, decisions to act pre-emptively are premised to a significant degree upon knowledge constructed by the deciding authority. The purity of such a decision is thus quite striking, since it involves not simply the weighing of existing evidence, but the active creation of knowledge about the future to serve as the evidentiary basis for interventions in the present. In other words, precisely because the unknown future being acted upon is something of an epistemic abyss, any decision to intervene pre-emptively 'becomes in the true sense absolute', to again quote Schmitt (Schmitt 2005: 12). Just as they suspend the juridical order by default, therefore, pre-emptive security practices also enact a decisionist form of political authority by what amounts to logical necessity. These two points suggest an originary conceptual relationship between pre-emption and exceptionalism, whose consequences for the organisation and exercise of political power, as well as experiences of political subjectivity, are profound and must be taken seriously if we are to adequately understand the contemporary global politics of (in) security.

Precarious Subjectivities

A particularly important consequence in this regard is that the exceptionalist politics presupposed by the logic of pre-emptive security brings into being a relationship between sovereign authorities and those governed thereby in which the latter are rendered continuously vulnerable to sudden and potentially violent interventions by the former. This is because the only effective limitations on the exercise of sovereign power in this context are the limits of the sovereign imagination itself. Indeed, relevant sovereign deciders are tasked with taming the future's radical contingency; and because the

---

pre-empting imagined sources of harm' (2008, 57). Prominent examples include exceedingly generous readings of entrapment provisions by the courts in cases where informants are used to goad suspects into hatching terror plots that are then pre-emptively foiled (Greenwald and Fishman 2015), or the European Union's expanded use of financial 'blacklisting' practices that significantly lower the evidentiary threshold for enacting 'targeted sanctions' against suspected terrorist financiers that effectively preclude their participation in any type of financial transaction (de Goede 2011, Sullivan and Hayes 2010).

epistemic basis for action towards this end will rely on imagined futures of the sovereign's construction, targets are identified based on subjective discretion rather than more objective juridico-normative guidelines. And since none can be summarily absolved of suspicion, all are always already constructed as possible targets for pre-emptive action (Ericson 2008). What thus emerges is a condition in which *any* individual may be subjected to what amounts to an arbitrarily decided anticipatory act at any time, such that regimes of (in) security governance premised upon a logic of pre-emption enact a decidedly precarious experience of political subjectivity.

To be sure, for almost all individuals, this perpetual vulnerability to potentially violent interventions based on conjecturally imagined futures will never be translated into an act of arbitrary sovereign violence. But the conceptual point being made here concerns the ever-present *potential* of this occurring, since even if no such action ever takes place, the continuous *possibility* that it will is a defining feature of the relationship between sovereign and subject under a pre-emptive security regime. And this possibility is a direct consequence of the enhanced decisional discretion vested in the sovereign by the logic of pre-emption, in that anticipatory decisions are placed outside the circumscriptions of the juridical order, thus removing the normative barriers that protect individuals from being targeted on the basis of speculative knowledge that they have no capacity to contest. The point, in other words, is that even if such arbitrary targeting never actually takes place, this will not be because of any normative constraints upon the sovereign's decisional authority; it will be because the sovereign decides against it.

The relations between sovereign and subject enacted by a logic of pre-emptive security in this sense closely resemble those that characterise the 'exceptional' political spaces theorised extensively in the Critical Security Studies literature. These are typified by what Agamben terms the 'camp,' which is characterised by sovereign power confronting its subjects 'without any mediation,'—meaning the sovereign's ability to act is unbound by legal norms while subjects are deprived of any agency to contest its decisions (Agamben 2000: 41; 1998: 171). Just as in the proverbial camp, therefore, the possibility for any individual to be arbitrarily subjected to violent sovereign interventions is always there under a pre-emptive security regime. Indeed, even in states ostensibly committed to the rule of law and human rights norms, the adoption of pre-emptive security strategies creates an ever-present *potential* for anyone to be inscribed as the sort of 'bare life' against which 'everything is possible' (Agamben 1998: 170).

The September 2011 targeted killing via drone strike of Al Qaeda operative and US citizen Anwar al-Awlaki provides a useful illustration of these considerations, as the incident highlights both the type of action made possible by the logics of pre-emptive security and the precarious subjectivities that are thereby enacted. The Obama administration's decision to target al-Awlaki can be read as an exercise in pre-emptive security, in that the broader 'killing programme' of which it was a high profile example has been framed in precisely such terms by its proponents (Leander 2011). In a 2012 interview, for instance, former National Counterterrorism Centre head Michael Leiter asserted that targeted killing was embraced by Obama as the most appropriate response to the 'situation where he is being told people *might* attack the United States *tomorrow*' (quoted in Becker and Shane 2012, emphasis added). This suggests the administration sees drone warfare as an effective strategy for ensuring that such catastrophic futures do not come to pass. Moreover—and with respect to the al-Awlaki case in particular—the pre-emptive character of the killing is highlighted by the administration's subsequent framing of the incident, whereby its legitimacy was affirmed by specifically invoking a pre-emptive imperative. The administration's initial response to critics thus asserted that al-Awlaki 'posed some sort of imminent threat', which justified such 'extraordinary measures' as the government killing a citizen without what would conventionally be understood as due process (Koring 2011).

Besides being an exemplar of pre-emptive security, however, the al-Awlaki case also represented a prototypically 'exceptional' act, as it embodied both components of political exceptionalism discussed above. Regarding the first—the suspension of the juridical order—a strong case can be made that the killing was extrajudicial, in that it was not authorised through established legal channels or in accordance with associated standards of evidence, and as such, was both ordered and carried out absent the due process of law constitutionally guaranteed to US citizens. The act can thus be understood as having suspended the legal order at the moment of its occurrence. Moreover, that targeting al-Awlaki in this way would contravene the juridical order seems to have been apparent to the Obama administration, as it sought to further justify the killing through claims of executive discretion on issues of national security. This can be seen in an internal memo from the Justice Department obtained by the *New York Times*, which was prepared with specific reference to the al-Awlaki case and asserted that 'while the Fifth Amendment's guarantee of due process applied, it could be satisfied by internal deliberations in the executive branch' alone (quoted in Becker and Shane 2012).

This circumventing of the juridical order hints at how the al-Awlaki case also embodies the second core component of exceptionalism—a decisionist paradigm of sovereign authority. In this respect, the decision to kill al-Awlaki—as with nearly all instances of drone strikes—was made through a by now well-established process that assigns such life-and-death prerogatives to a highly select group of officials within the executive branch. This process was outlined in a controversial *New York Times* investigative report published in 2012, which is worth quoting at length to illustrate the degree to which the exercise of sovereign authority in the context of drone warfare takes a distinctly decisionist form:

> Every week or so, more than 100 members of the government's sprawling national security apparatus gather, by secure video teleconference, to pore over terrorist suspects' biographies and recommend to the president who should be the next to die. This secret 'nominations' process is an invention of the Obama administration, a grim debating society that vets the PowerPoint slides bearing the names, aliases and life stories of suspected members of Al Qaeda's branch in Yemen or its allies in Somalia's Shabab militia (*sic*)... [N]ames go off the list if a suspect no longer appears to pose an imminent threat...The nominations [then] go to the White House, where by his own insistence and guided by [chief counterterrorism advisor Jim] Brennan, Mr. Obama must approve any name (Becker and Shane 2012).

This description highlights how the final authority to decide who is to be killed and when is granted directly to the president alone. Thus, starting from the assumption that executive branch deliberation followed by the president signing off constitutes due process, the executive is freed from any normative circumscriptions regarding this use of violence in this way. This vests within the person of the president the discretionary capacity to determine who to target, when to strike, and what counts as adequate evidence that someone poses a sufficient threat to be killed. Such prerogatives conspicuously mirror those ascribed to the decisionist sovereign under a politics of exceptionalism.

By thus illustrating how the anticipatory exercise of sovereign power shifts relations between sovereign and subject towards an effectively unmediated confrontation, the al-Awlaki case highlights both the originary relationship between pre-emption and exceptionalism, and the implications of this link for the character of political subjectivity under a politics of pre-emptive security. Indeed, once the president made the pre-emptive decision to target al-Awlaki, the latter could immediately be killed with impunity by the agents of American sovereignty. The law thus no longer served as an effective mediator between sovereign and subject, since despite being a US citizen, al-Awlaki could still be killed on the basis of what amounted to an executive decree. The juridico-normative limitations on the president's decisional authority were therefore subordinated to an imperative to govern the future with which such limits are in many ways incompatible. When faced with a sovereign power that could unilaterally decide when he would be killed and against which he had no practical recourse, al-Awlaki was therefore constructed as precisely the sort of 'bare life' that Agamben and others associate with the exceptionalism of the 'camp' without actually being located in the sort of explicitly defined space with which this term is more often associated (Agamben 2000). The al-Awlaki killing thus provides an instructive example of the sort of exceptional practices that are made possible by the logic of pre-emptive security. The crucial point in this regard is that, under the pre-emptive security regime being prosecuted through the American drone programme, if President Obama had decided *not* to kill al-Awlaki, this would not have been because of any perceived legal limits upon his doing so; it would have instead been because the president himself simply *decided* not to.

To be sure, the heated debate over the legitimacy of the al-Awlaki killing suggests that alternative understandings of the case and its implications are also compelling. For instance, al-Awlaki's propagandistic activities on behalf of Al Qaeda—with whom the US has been engaged in a 'war' since 9/11— might be understood as rendering the killing neither pre-emptive in nature— since it can be viewed merely as an operation against a combatant during ongoing hostilities—nor extrajudicial—since al-Awlaki can be seen to have forfeited his citizenship protections by actively aiding an enemy force on foreign soil. These considerations reflect the Obama's administration's views and must be taken seriously in any discussion of the strike's legitimacy. Yet regardless of one's position on this normative question, the incident still illustrates the key conceptual point being made in this section—namely, that the imperative to tame temporal contingency conflicts with key normative mediations between sovereign and subject in a liberal democracy. Indeed, the very fact that the al-Awlaki killing could take place at all suggests that a state of perpetual vulnerability to anticipatory sovereign violence is a defining feature of political subjectivity under a pre-emptive politics of (in)security, even if subjects are never actually targeted in practice. Again, this condition is a result of the *de facto* elimination of juridico-normative mediation between sovereign power and political subjects, which in turn stems from a prioritisation of the imagination that is itself a function of the imperative to tame a radically uncertain future. The creation of such precarious subjective conditions is thus precisely what is at stake with the implementation of pre-emptive security strategies—a point that must be taken seriously whenever the adoption of such strategies is being considered.

Conclusion

What I have termed the 'temporalisation' of (in)security represents merely one example of a broader trend in societal governance characterised by a 'shift from responding to past events to preventing future harms' (Dershowitz 2006: 7). Evidence of this shift can be found across myriad areas of human affairs, as sectors as diverse as financial regulation (Porter 2009), public health management (Cooper 2006), crime prevention (Ericson 2007), urban planning (Coaffee 2009), and natural disaster management (United Nations 2013) are increasingly characterised by strategies, technologies, and rationalities based on anticipatory logics. The arguments developed above suggest that framing key problems of societal governance in this way can pose significant challenges to the constitutive norms of the liberal democratic polities often in the vanguard of this trend. Indeed, the exceptionalist politics that I have argued are presupposed by the temporalities of pre-emption in many ways run counter to such basic precepts as the rule of law and associated limitations on the prerogatives of sovereign authority. Yet this need not to imply that a politics of pre-emption can never constitute a legitimate approach to a particular problem, as the radical contingencies of late modernity may leave few other options (Beck 2008). The point is rather to emphasise that only when the sorts of practices made possible by the underlying logics of anticipatory governance are adequately considered can decisions to implement practical strategies based thereupon be made responsibly. And in an era when the effective governance of a political space is increasingly reliant upon a capacity to govern the unfolding of time, such questions are perhaps more 'timely' than ever.

References

Aalberts, T. and Werner, W. (2011) 'Mobilising Uncertainty and the Making of Responsible Sovereigns.' *Review of International Studies* 37: 2183-2200.

Ackerman, S. (2015) 'No-Fly List Uses 'Predictive Assessments' Instead of Hard Evidence, US Admits.' *The Guardian*. 10 August. Available online at: http://www.theguardian.com/us-news/2015/aug/10/us-no-fly-list-predictive-assessments

Agamben G. (1998) *Homo Sacer: Sovereign Power and Bare Life*. Stanford CA: Stanford University Press.

Agamben, G. (2000) *Means Without End: Notes on Politics*. Minneapolis: University of Minnesota Press.

Agamben, G. (2005) *State of Exception*. Chicago: University of Chicago Press.

Amoore, L. (2014) *Politics of Possibility: Risk and Security Beyond Probability.* Durham, NC: Duke University Press.

Anderson, B. (2010a) 'Pre-emption, Precaution, Preparedness: Anticipatory Action and Future Geographies.' *Progress in Human Geography* 34 (6): 777-798.

Anderson, B. (2010b) 'Security and the Future: Anticipating the Event of Terror.' *Geoforum* 41: 227-35.

Aradau, C. and van Munster, R. (2007) 'Governing Terrorism Through Risk: Taking Precautions, (Un)Knowing the Future.' *European Journal of International Relations* 13 (1): 89-115.

Aradau, C. and van Munster, R. (2008) 'Taming the Future: The *Dispositif* of Risk in the War on Terror.' In: Amoore, L. and de Goede, M. (eds) *Risk and the War on Terror*. London: Routledge: 23-40.

Aradau, C. and van Munster, R. (2009) 'Exceptionalism and the "War on Terror": Criminology Meets International Relations.' *British Journal of Criminology* 49: 686-701.

Aradau, C. and van Munster, R. (2011) *Politics of Catastrophe: Genealogies of the Unknown*. London: Routledge.

Beck, U. (2008) *World at Risk*. Cambridge: Polity Press.

Becker, J. and Shane, S. (2012) 'A Measure of Change: Secret 'Kill List' Proves a Test of Obama's Principles and Will.' *The New York Times*. 29 May. Available online at: http://www.nytimes.com/2012/05/29/world/obamas-leadership-in-war-on-al-qaeda.html?_r=2&

Coaffee, J. (2009) *Terrorism, Risk, and the Global City*. London: Ashgate.

Cooper, M. (2006) 'Pre-empting Emergence: The Biological Turn in the War on Terror.' *Theory, Culture & Society* 23(4): 113-135.

de Goede, M. (2012). *Speculative Security: The Politics of Pursuing Terrorist Monies*. Minneapolis: University of Minnesota Press.

de Goede, M. (2011) 'Blacklisting and the Ban: Contesting Targeted Sanctions in Europe.' *Security Dialogue* 42(6): 499–515.

de Goede, M and Randalls, S. (2009) 'Precaution, Pre-emption: Arts and Technologies of the Actionable Future.' *Environment and Planning D: Society and Space* 27: 859-78.

Dershowitz, A. (2006) *Pre-emption: A Knife that Cuts Both Ways*. New York: Norton.

Elmer, G. and Opel, A. (2006) 'Surviving the Inevitable Future: Pre-emption in an Age of Faulty Intelligence.' *Cultural Studies* 20(4-5): 477-92.

Ericson, R. (2007) *Crime in an Insecure World*. Cambridge: Polity Press.

Ericson, R.V. (2008) 'The State of Pre-emption: Managing Terrorism Risk Through Counter Law.' In: Amoore, L. and de Goede, M. (eds) *Risk and the War on Terror*. London: Routledge. 57-76.

Government of Canada (2011) *Building Resilience Against Terrorism: Canada's Counter-Terrorism Strategy*. Ottawa: Government of Canada.

Greenwald, G. and Fishman, A. (2015) 'Latest FBI Claim of Disrupted Terror Plot Deserves Much Scrutiny and Skepticism.' *The Intercept*. January 16. Available online at: https://firstlook.org/theintercept/2015/01/16/latest-fbi-boast-disrupting-terror-u-s-plot-deserves-scrutiny-skepticism/

Herman, S. (2011) *Taking Liberties: The War on Terror and the Erosion of American Democracy*. Oxford: Oxford University Press.

Humphreys, A. (2014) 'Canadian "Terror Tourist" Mohamed Hersi Gets 10 Years in Jail for Planning to Join Islamic Jihadist Group in Somalia.' *National Post*. July 24. Available online at: http://news.nationalpost.com/news/canada/canadian-terror-tourist-mohamed-hersi-gets-10-years-in-jail-for-planning-to-join-islamic-jihadist-group-in-somalia

Huysmans, J. (2004) 'Minding Exceptions: The Politics of Insecurity and Liberal Democracy.' *Contemporary Political Theory* 3: 321-41.

Isin, E. and Rygiel, K. (2007) 'Abject Spaces: Frontiers, Zones, Camps.' In: Dauphinée, E. and Masters, C. (eds) *The Logics of Biopower and the War on Terror: Living, Dying Surviving*. Houndmills: Palgrave-MacMillan: 181-203.

Kessler, O. (2011) 'The Same as it Never Was? Uncertainty and the Changing Contours of International Law.' *Review of International Studies* 37: 2163-82.

Kessler, Ol and Daase, C. (2008) 'From Insecurity to Uncertainty: Risk and the Paradox of Security Politics.' *Alternatives* 33: 211-232.

Koring, P. (2011) 'Memo Equips Obama with "Licence to Kill."' *The Globe and Mail*. 9 October. Available online at: http://www.theglobeandmail.com/news/world/americas/memo-equips-obama-with-licence-to-kill/article2196145/

Lazar, N.C. (2006) 'Must Exceptions Prove the Rule? An Angle on Emergency Government in the History of Political Thought.' *Politics & Society* 34(2): 245-75.

Leander, A. (2011) 'Risk and the Fabrication of Apolitical, Unaccountable Security Markets: The Case of the CIA "Killing Program."' *Review of International Studies* 37: 2253-68.

Machiavelli, N. (1995) *The Prince*. Translated by Wootton, D. Indianapolis: Hackett.

Massumi, B. (2007) 'Potential Politics and the Primacy of Pre-emption.' *Theory & Event* 10(2).

Muller, B. (2010) *Security, Risk, and the Biometric State: Governing Borders and Bodies*. London: Routledge.

Opitz, S. and Tellmann, U. (2015) 'Future Emergencies: Temporal Politics in Law and Economy.' *Theory, Culture & Society* 32(2): 107-129.

Porter, T. (2009) 'Risk Models and Transnational Governance in the Global Financial Crisis: The Cases of Basel II and Credit Rating Agencies.' In: Helleiner, E., Pagliari, S. and Zimmermann, H. (eds) *Global Finance in Crisis: The Politics of International Regulatory Change*. London: Routledge. 56-73.

Schmitt, C. (2005) *Political Theology: Four Chapters on the Concept of Sovereignty*, Translated by G. Schwab. Chicago: University of Chicago Press.

Stockdale, L.P.D. (2013) 'Imagined Futures and Exceptional Presents: A Conceptual Critique of "Pre-emptive Security."' *Global Change, Peace & Security* 25(2): 141-57.

Sullivan, G. and Hayes, B. (2010) *Blacklisted: Targeted Sanctions, Pre-emptive Security, and Human Rights*. Berlin: European Centre for Constitutional and Human Rights.

Svendsen, A. (2010) 'Re-fashioning Risk: Comparing UK, US, and Canadian Security and Intelligence Efforts Against Terrorism.' *Defence Studies* 10(3): 307-335

Teotonio, I. (2011) "U of T Graduate's Arrest on Terror Charges Alarms Toronto Somalis." *Toronto Star*. March 31. Available online at: http://www.thestar.com/news/crime/article/965595--toronto-man-arrested-on-terrorism-related-charges

United Nations (2013) *Disaster Risk Reduction in the United Nations*. New York: United Nations Office for Disaster Risk Reduction.

Virilio, P. (2010) *The University of Disaster*. Cambridge: Polity Press.

Weber, C. (2007) 'Securitising the Unconscious: The Bush Doctrine of Pre-emption and *Minority Report*.' In: Dauphinée, E. and Masters, C. (eds) *The Logics of Biopower and the War on Terror: Living, Dying Surviving*. Houndmills: Palgrave-MacMillan: 109-28.

Conclusion

How Time Shapes our Understanding of Global Politics

CAROLINE HOLMQVIST
UNIVERSITÉ LIBRE DE BRUXELLES, BELGIUM AND SWEDISH
INSTITUTE OF INTERNATIONAL AFFAIRS, SWEDEN
&
TOM LUNDBORG
SWEDISH INSTITUTE OF INTERNATIONAL AFFAIRS AND
SWEDISH DEFENCE UNIVERSITY, SWEDEN

Making sense of politics in the modern world has always involved constructing conceptions of the 'past', 'present' and 'future' (Koselleck 2004 [1985]; Toulmin 1992). In other words, as Kimberly Hutchings tells us, assumptions of time 'fundamentally shape what we can and cannot know about world politics today' (Hutchings 2008). Building on recent theoretical insights into the significance of time for thinking about politics (Chambers 2003, Grosz 2005, Hutchings 2008, Widder 2008, Shapiro 2010, Lundborg 2012), this chapter shows how questions of time and politics can be variously posed and explored. Rather than presenting a central argument, we seek to open up a number of different directions that research into time and global politics may take. This discussion is structured around four empirical areas, which can be thought of as potential case studies. The first explores competing temporalities embodied in the United Nations (UN) system; the second looks at apocalyptic visions of the global climate crisis; the third focuses on the instantaneity of global information flows; and the fourth examines practices of pre-emption in global counterterrorism measures. In each case, the dominant role of teleological understandings of time and history in modern political discourse is questioned, and at least four conceptions of time other than the teleological invoked: cosmological time, eschatological time, instantaneous time, and time as a flow of becoming.

By tracing the competing temporalities at work in these given examples of world political affairs, this chapter provides an overview of the types of inquiry that a focus on time in global politics might generate. They serve to illustrate the manifold ways in which questions around time and politics can be pondered and explored – mindful of the way in which certain discourses of global politics are permitted, promoted and excluded in various ways by and through different conceptions of time. As such, this quartet of 'cases' testify to the competing modes not only of explaining and understanding, but also practicing world politics, enabled by different conceptions of time. In other words, they point to the politics of time in global affairs. Before turning to these potential studies we begin by giving an overview of how time has been used as a point of entry into studies of global politics, as well as provide a brief introduction to the alternative notions of time that we use in each 'case' by contrasting them with the dominant linear and teleological time associated with modern political discourse.

Time, Politics and Globalisation

The most cited books on contemporary globalisation are based on a rather limited engagement with time and temporality, which mainly focuses on the significance of speed, acceleration, and the compression of time and space. Much of this literature is based on David Harvey's (1990) concept of the 'time-space compression', which refers to how the ever-increasing speed of 'transnational flows' has collapsed distances in time as well as in space. The flows of, for example, money and information have become *instantaneous*, rendering spatial distances obsolete (Appadurai 1996, Urry 2003). The narrow focus on the significance of speed, acceleration, and the time-space compression is also evident in the literature more specifically concerned with the *political* dimension of contemporary globalisation. In this context, globalisation is mainly discussed by pointing to how the acceleration and speed of transnational flows challenge the static borders of sovereign territory (Der Derian 1992, Agnew 1994, Tuathail 1996, Scholte 2005, Held and McGrew 2007).

While the globalisation literature has opened up important questions about the potential significance of time for thinking about politics, there is much work to be done in order to advance our understanding of how *different* conceptions of time can be used for examining, analysing and understanding global politics and political life under globalizing conditions. In order to do this, a more expansive and nuanced analysis of time and its relation to politics is necessary. Therefore, we need other adjacent fields of study, which can be used in order to investigate the significance of time as a constitutive dimension of politics.

On the one hand, there is a growing interest in time and temporality among philosophers, social theorists, and scholars of International Relations (IR) (Hoy 2009, Hom 2010, Ruin and Ers 2011). On the other hand, there is a literature more specifically concerned with the *politics* of time and temporality. For example, in his book *Untimely Politics* Samuel Chambers (2003) has challenged the dominant linear conception of time and argued for an alternative non-linear way of thinking about politics, mainly by drawing on the work of Nietzsche, Foucault and Derrida. Elizabeth Grosz has also problematised the linear conception of time in her book *Time Travels: Feminism, Nature, Politics* (2005), which analyses the significance of time as a productive force that conditions different forms of political struggles, in particular feminist struggles. Kimberly Hutchings' work on *Time and World Politics* (2008) was mentioned earlier and stands as the most significant theoretical contribution to thinking about the relationship between theories of time and theories of world politics. Another important theoretical contribution is Nathan Widder's *Reflections on Time and Politics* (2008), which offers a series of critical reflections on the limits of the linear progressive view of time and opens up to a more radical understanding of time and politics based on continental and poststructural philosophy. In a more empirical sense, Michael J. Shapiro's book *The Time of the City* (2010) addresses the politics of urban life through the lens of various philosophical conceptions of time. Finally, Tom Lundborg's book entitled *Politics of the Event: Time, Movement, Becoming* (2012) demonstrates how the concept of the 'event' can be used for analysing a sovereign politics of time seeking to reproduce ideas about a 'modern' political order through the inscription of 'borders in time' separating the past from the present and the future. Like many of these works we are interested in challenges to the dominant teleological concept of time.

Alternative Concepts of Time

According to the dominant teleological conception of time, past, present and future are conjoined linearly, suggesting a 'forward movement' or 'progress'. The linear understanding of time is strongly associated with the grand ideologies that dominated the nineteenth and twentieth centuries: liberalism and communism. With the end of the Cold War came the declaration that history itself had come to an end (Fukuyama 1992). While Fukuyama's thesis has been widely criticised, the teleological conception of time it relies upon continues to have formidable influence, not least on discourses of global economic development that express confidence in continuous growth and the promise of progress from 'under-developed' to 'developed'.

At least four additional conceptions of time to the teleological can be identified in the tradition of intellectual history. First, the **cosmological** conception of time derives from Aristotelian thinking and was further developed in

Renaissance philosophy via early modern science and entails a view of time as *cyclical* and *repetitive* (Cassirer 1979). This view of time has for example been used to think about the repetitive nature of politics in the modern era, and the eternal movements of inter-state politics in a world that lacks a higher authority than the sovereign state (Walker 2010).

Second, the **eschatological** conception of time is concerned with time as finite and the world coming to an end, and is strongly associated with theological Judaic-Christian traditions (Löwith 1949, Taubes 2009 [1947]). This view of time has for example been used to think about the values and fears that often underlie ideas about the world suddenly coming to an end via nuclear war, terrorist attacks or environmental disasters (Bradley and Fletcher 2010).

Third, the conception of **instantaneous time** relates to the view of the contemporary era as one of 'instantaneity', characterised by the time-space compression and the ways in which time has become 'radically present' (Harvey 1990, Beck 1992, Heller 1999, Bauman 2000). This notion of time has for example been used to highlight the implications of the rapid increase of transnational flows – of people, money and information – which some see great potential in, while others point to the fears, dangers and risks commonly associated with them (Lundborg 2011).

Fourth, the conception of time as a **flow of becoming** was developed by 20th Century continental philosophers Gilles Deleuze and Jacques Lacan, among others, and points to how time eludes the static being of the individual subject, prompting attempts to construct fantasies of the full presence of the coherent subject (Deleuze 1994, Lacan 2006). This view of time has for example been drawn upon to study a politics seeking to construct fantasies of the full presence of the individual subject *within* the limits of the sovereign territorial state, but also to express different modes of resistance that try to elude the authority of the state (Edkins 2011, Lundborg 2012).

We believe that some of the most hotly debated issues in contemporary global politics – including the global climate crises, global counterterrorism measures, global information flows, and the limits and possibilities of progress in the international order – can be rearticulated, rethought and analysed in a radically new light via a focus on conceptions of time. In this respect, while the dominant teleological view can be linked to a particular understanding of political life based on liberal ideals of freedom, progress and enlightenment, the other conceptions of time highlight alternative understandings of political life.

For example, discourses based on cosmological time can be linked to a form of political life that is reluctant to change and aims to continue, eternally, along the same road. Discourses based on eschatological time might instead point to a form of political life dominated by the constant fear of an abrupt end through, for example, apocalypse or disaster. In turn, discourses based on instantaneous time open up the possibility of new forms of political belonging through instantaneous connections between people located in different parts of the world. Moreover, time as a contingent flow of becoming can be linked to a more radical understanding of life, whereby life constantly eludes the full presence and completeness of the subject. To demonstrate how this may be accomplished we now turn to our four suggested case studies.

Studying Time and Global politics

Since our specific focus is on the relationship between concepts of time and global politics, we have chosen four suggested case studies that all point to a distinctively global or transnational dimension of contemporary politics. In other words, they all relate to that which *transcends* or happens *beyond* the limits of the sovereign territorial state. In this respect, different concepts of time can be used both in order to take us beyond the modern state as well as to point to the various tensions between what lies beyond and what lies within the state/state system; tensions that emerge when different concepts of time clash with one another.

I: Competing temporalities in the United Nations system

Coming to terms with the international and what may be said to constitute its 'order' is perhaps the main preoccupation of the discipline of IR. While the category of space has dominated research into the international and the question of order, the category of time remains under-investigated and under-explored. This is noteworthy: questions of how we think of change on the international level, of *stability* contra *rupture*, of *possibility* contra *determinism*, of duration and longevity versus finitude, and so on – all such questions are conditioned by understandings of time and temporality. This is abundantly clear when we turn to the organisation that more than any other can be said to embody the attempted construction of international order in the present day: the UN.

The inquiry into the competing temporalities at work in the construction of international order can be approached via a study of the UN as organisation. One way to structure such investigation would be to depart from the plethora of documents generated by the UN: from the UN Charter, key Security Council resolutions, United Nations Development Programme (UNDP) reports to speeches by the UN Secretary General. These documents as texts neatly encompass the on-going and necessarily incomplete construction of international order. By focusing on the sui generis global organisation, the UN, we expect that such empirical research would yield insight into some of the central themes of mainstream global political discourse, in particular discourses of security and social and economic development.

The international order that is embodied in the UN system is at heart a liberal order, vacillating between, on the one hand, the promotion of the 'progress' of individual human life and, on the other, the fundamental dependence on the active support and participation of individual sovereign states, with the attendant principles of territorial integrity and non-intervention (Chandler 2002, Cunliffe 2011, Bellamy 2011 and 2005). Thus the investigation of the temporalities of the international à travers the temporalities embodied in the UN system is at once an investigation of the time of the international and at once an investigation of the competing temporalities of liberal international theory. While the teleological conception of time as everlasting progress is most strongly associated with liberal thought, we can expect this trajectory to be juxtaposed with a cosmological view of time as dominated by eternal cycles and repetition. In focusing on this interaction, the study of the political logics embedded in the UN system would shed new light on the limits of 'progress' in a world governed by the eternal movements of inter-state politics. The conception of time as instantaneous would offer additional perspective on the workings of the international system as embodied in an organisation that struggles to deal with the long-term politics of poverty reduction, climate change and so on, incessantly challenged by states' reactive concern for the immediate.

II: Apocalyptic visions of the global climate crisis

A second potential area of interest concerns the global climate crisis. It is well established that one of the root causes of this crisis is the continuous aspiration for increased productivity/economic growth and its detrimental effects on the environment and the sustainability of the earth's ecosystems (Helm and Hepburn 2009). Thus, there is an important tension to be examined here: between the modern belief in the individual's as well as the state's right to continuous progress on one hand, and on the other hand an apocalyptic vision of the world, based on an eschatological conception of time as finite and the world rapidly coming to an end.

The tensions created between linear and eschatological time in the context of the global climate crisis can certainly be expressed in different ways depending on the discourse one is looking at. While the linear notion of time might be more dominant in discourses of national security, for example, eschatological time is often expressed in popular culture including films such as *The Day After Tomorrow* (2004), *An Inconvenient Truth* (2006), and *Melancholia* (2011).

While studying the ways in which different discourses give priority to distinct notions of time, it would also be fruitful to look more closely at the interactions between eschatological and linear time. Another example of how this may be done is to study UN climate summits, in which clashes between opposing views of the climate crisis often become apparent. While some states highlight the extreme dangers of the climate crisis and point to the risks of inaction, others merely state that it is not in their national interest to take action in the present.

When studying the interaction between eschatological and linear notions of time we may also ask how they condition – rather than simply exclude – one another. For example, it could be argued that linear progressive time relies on the notion of a final end, which gives the search for continuous progress purpose and meaning in the first place, just like death gives meaning to life. On that basis, it is possible to investigate how fantasies of an absolute end of the world do not automatically threaten the idea of continuous progress of states and individuals but supplies the latter with its necessary and underlying condition: namely the constant risk of coming to an end. To progress then implies a contingent and unpredictable process of becoming, which is 'meaningful' only as long as it may suddenly come to an end. Hence the concept of time as a flow of becoming is also relevant to include in this analysis – not in order to think of becoming as 'eternal', but to highlight an unpredictable movement based on elements of chance and contingency. Indeed, it can be argued that it is precisely our inability to fully predict the future that gives meaning to life, and which makes all attempts to calculate what *will* happen futile. Eschatological time can thus be seen as a condition of possibility for linear progressive time as it poses a constant threat to linear time of reaching a sudden end.

III: The instantaneity of global information flows

Investigation into the subject of global information flows presents a third potential avenue for investigation of the temporality of the global political domain. Recent decades have seen a surge in interest in how the spread of information and communication between individuals and corporate entities contributes to a distinct media sphere or global network society (Castells 2010). While some commentators see real political potential in social networking (often citing the 'Arab Spring'), more sinister manifestations of global information flows can be found in the increased reliance on global surveillance techniques and the military use of remote killing through unmanned aerial vehicles. Hence there is striking contrast between the diagnoses of the political potential of temporal instantaneity: promise on the one hand and risks and dangers on the other (Lundborg 2011). As social theorists have asserted for some time: delving more deeply into the constitutive role of speed, proximity, acceleration (Virilio 2006, Rosa, 2013) and instantaneity for global political discourse is imperative for understanding the limits and potentials of political action today.

Exploring the paradoxical effects of the instantaneity of information flows requires attention to the active role of materiality in global surveillance techniques and its importance for the military use of remote killing (Gregory 2011, Wall and Monihan 2011, Plaw and Fricker 2012). Previous research into the force of materiality in social and political relations (Lundborg and Vaughan-Williams 2011 and 2015, Holmqvist 2013) points to the 'living' potential of material force in political affairs; and the assemblage structure of contemporary political phenomenon (Latour 2005, Bennett 2010). From the range of empirical materials that may be selected as basis for such investigation, we have chosen to highlight elements pertaining to global violence, and the impact of violent materialities on how politics is (or can be) enacted. Here, testimonials from drone operators, reports from the locations where drone operations take place, and official documents from the United States' defence and security establishment outlining the (current and future) use of unmanned weapons systems are noteworthy.

The contrast between and clash of teleological and instantaneous conceptions of time are expected to be of primary significance in this area of research. While the wars of (classical) modernity were imbued with teleological conceptions of time – evident in the conception of wars as temporally demarcated, bracketed by dates of beginning and end – wars conditioned by instantaneous time oscillate between a radical sort of present and endlessness. The breakdown of the distinction between 'war' and 'peace' in our time – indeed the breakdown between the *finite* and the *infinite* in war – demands new political temporalities. Given that the surveillance-wars never 'start' they can never 'end' in any conventional sense of the word either: reckoning with this is both theoretically and politically challenging.

IV: The politics of preemption in global counterterrorism policy and practice

The fourth case is linked to the politics of preemption in global counterterrorism policy and practice. The concept of preemption refers to the strategy of pre-empting future attacks before they materialise (Rasmussen

2006, Massumi 2007). Following the discourses of the Global War on Terror, this strategy has been discussed extensively in recent years, with particular reference to the idea of protecting individual freedoms and liberties by making *exceptions* to international as well as domestic laws and norms (Neal 2010).

By highlighting the conception of time as a contingent flow of becoming we are able to add another layer to this topic by looking at how the politics of preemption can be examined and analysed in relation to fantasies of controllability. In this way, we can also shed light on some of the most controversial elements of contemporary counterterrorism measures, including indefinite detention, extraordinary rendition, the use of torture and secret prison facilities. Examples of empirical material to be examined include recent human rights-reports from organisations like Human Rights Watch and Amnesty International, as well as official documents from the US Departments of Defense and Homeland Security. Moreover, one may study the cultural expressions of these practices in films and TV-series like *Minority Report* (2002), *Rendition* (2007), *24* (2001-2010), and *Homeland* (2011-present).

While much of this discourse seems to suggest that contemporary political life under globalising conditions is increasingly unpredictable, it also expresses a certain desire to control the unpredictable. The challenge of satisfying this desire lies precisely in how the flow of becoming renders impossible the existence of a stable and present 'now', from which we can safely observe what has happened in the past and make predictions about the future. The unpredictability of terrorist attacks and the invention of new methods with which they are carried out are thus responded to with new ways of dealing with the contingent flows of becoming. Crucially, much of this discourse seeks to respond to the 'futurity' of becoming by trying to imagine the 'unimaginable' of what might happen next. In this context, therefore, we need to study how counterterrorism policy and practice express a certain politics of time that prioritises the future over the present, and indeed appears to bring the future into the present in a remarkable non-linear fashion. The idea of acting on the future in the present brings us back to instantaneous time as it points to the ways in which policies are enacted *immediately*, without waiting for events to unfold. What needs to be studied, thus, is the interaction between the unpredictable flow of becoming and the instantaneity of the sovereign decision.

Conclusion

By using time as a primary analytical lens, scholars will be able to rethink and rearticulate the theoretical frameworks through which global political discourses are examined, understood and analysed; provide new empirical insights into the conditions that shape political life in the contemporary world; and shed new light on our capacity to think about the limits and possibilities of 'change' in global politics. This approach holds the promise of generating a series of inquiries, as part of new research agenda with the potential for both theoretical innovation and substantive contributions to current political debates. Perhaps most significantly, this kind of agenda would take us beyond the narrow concern with progressive, linear and teleological time that has dominated the ways in which time is understood as mere 'history' in the discipline of IR. Any attempt to come to terms with the profound challenges of global politics must look beyond that narrow understanding of time and open up an expanded view of the multiple temporalities shaping political life in the contemporary era.

References

Agnew, J. (1994) 'The Territorial Trap: The Geographical Assumptions of International Relations Theory.' *Review of International Political Economy* 1(1): 53-80.

Appadurai, A. (1996) *Modernity at Large: Cultural Dimensions of Globalisation*. Minneapolis: University of Minnesota Press.

Ballard, J. G. (1964) *The Burning World*. Berkley Books.

Bauman, Z. (2000) *Liquid Modernity*. Cambridge: Polity.

Beck, U. (1992) *Risk Society: Towards a New Modernity*. London: SAGE.

Bellamy, A. (2005) *International Society and its Critics.* Oxford: Oxford University Press.

Bellamy, A. (2011) *Global Politics and the Responsibility to Protect.* London: Routledge.

Bennett, J. (2010) *Vibrant Matter: a Political Ecology of Things* Durham and London: Duke University Press.

Bradley, A. and Fletcher, P. (eds) (2010) *The Politics to Come: Power, Modernity and the Messianic*. London: Continuum.

Cassirer, E. (1979) *The Individual and the cosmos in Renaissance Philosophy*. [New ed.]. 2. pr. Philadelphia: University of Pennsylvania Press.

Castells, M. (2010) *The Rise of the Network Society*. Second edition. Chichester, West Sussex: Wiley-Blackwell.

Chambers, S. (2003) *Untimely Politics*. Edinburgh: Edinburgh University Press.

Chandler, D. (2002) *From Kosovo to Kabul: Human Rights and International Intervention*. London: Pluto.

Cunliffe, P. (ed.) (2011) *Critical Perspectives on The Responsibility to Protect: Interrogating Theory and Practice*. London: Routledge.

Deleuze, G. (1994) *Difference and Repetition*. London: Athlone.

Der Derian, J. (1992) *Antidiplomacy: Spies, Terror, Speed, and War*. Cambridge, Mass.: Blackwell.

Edkins, J. (2011) *Missing: Persons and Politics*. Ithaca: Cornell University Press.

Foucault, M. (2006 [1964]) *Madness and Civilisation: A History of Insanity in the Age of Reason*. Translated by Howard, R. London: Routledge.

Foucault, M. (2002 [1969]) *The Archeology of Knowledge*. Translated by Sheridan Smith, A. M. London: Routledge.

Fukuyama, F. (1992) *The End of History and the Last Man*. London: Hamish Hamilton.

Gregory, D. (2011) 'From a View to Kill: Drones and Late Modern War.' *Theory, Culture and Society* 27(7-8): 190-194.

Grosz, E. (2005) *Time Travels: Feminism, Nature, Power*. Durham: Duke University Press.

Harvey, D. (1990) *The Condition of Postmodernity: An Enquiry Into the Origins of Cultural Change*. Oxford: Blackwell.

Held, D. and McGrew, A. (2007) *Globalisation/Anti-Globalisation*. Second edition. Cambridge: Polity.

Heller, A. (1999) *A Theory of Modernity*. Malden, Mass.: Blackwell Publishers.

Helm, D. and Hepburn, C. (eds) (2009) *The Economics and Politics of Climate Change*. Oxford: Oxford University Press.

Holmqvist, C. (2013) 'Undoing War: War Ontologies and the Materiality of Drone warfare.' *Millennium Journal of International Affairs* 41(3).

Holmqvist, C. (2010) *Policing Wars: a Twenty-First Century Discourse on War*. PhD Thesis, Department of War Studies. King's College London.

Holmqvist, C. (2010) 'Perpetual Policing Wars.' In: Holmqvist, C. and Coker, C. (eds) *The Character of War in the Twenty-First Century*. London: Routledge.

Hom, A. R. (2010) 'Hegemonic metronome: the ascendancy of Western standard time.' *Review of International Studies* 36(4): 1145-70.

Hoy, D.C. (2009) *The Time of our Lives: A Critical History of Temporality*. Cambridge, MA: MIT Press.

Hutchings, K. (2008) *Time and World Politics: Thinking the Present*. Manchester: Manchester University Press.

Koselleck, R. (2004 [1985]) *Futures Past: on the Semantics of Historical Time*. New York: Columbia University Press.

Lacan, J. (2006) *Ecrits: The first Complete Edition in English*. New York: W.W. Norton & Co.

Latour, B. (2005) *Reassembling the Social: An Introduction to Actor-Network-Theory*. Oxford: Oxford University Press.

Lundborg, T. (2012) *Politics of the Event: Time, Movement, Becoming*. London: Routledge.

Lundborg, T. (2011) 'What Lies Beyond Lies Within: Global Information Flows and the Politics of the State/Inter-state System.' *Alternatives: Global, Local, Political* 36(2): 103-117.

Lundborg, T. and Vaughan-Williams, N. (2015) 'New Materialisms, Discourse Analysis, and International Relations: a Radical Intertextual Approach.' *Review of International Studies* 41(1): 3-25.

Lundborg, T. and Vaughan-Williams, N. (2011) 'Critical Infrastructure, Resilience, and Molecular Security: the Excess of "life" in Biopolitics.' *International Political Sociology* 5(4): 367-383.

Löwith, K. (1999 [1949]) *Meaning in History*. Chicago: University of Chicago Press.

Massumi, B. (2007) 'Potential Politics and the Primacy of Preemption.' *Theory & Event* 10(2).

McCarthy, C. (2007). *The Road*. New edition. London: Picador.

Neal, A.W. (2010) *Exceptionalism and the Politics of Counter-terrorism: Liberty, Security and the War on Terror*. London: Routledge.

Plaw, A. and Fricker, M.S. (2012) 'Tracking the Predators: Evaluating the US Drone Campaign in Pakistan.' *International Studies Perspective* 13(4): 344-365.

Rasmussen, M.V. (2006) *The Risk Society at War: Terror, Technology and Strategy in the Twenty-First Century*. Cambridge: Cambridge University Press.

Ruin, H. & Andrus, E. (eds) (2011) *Rethinking Time: Essays on History, Memory, and Representation*. Huddinge: Södertörns högskola.

Scholte, J.A. (2005) *Globalisation: A Critical Introduction*. Second edition. New York: Palgrave Macmillan.

Shapiro, M.J. (2010) *The Time of the City: Politics, Philosophy and Genre*. London: Routledge.

Taubes, J. (2009 [1947]) *Occidental Eschatology*. Stanford: Stanford University Press.

Toulmin, S. (1992) *Cosmopolis: The Hidden Agenda of Modernity*. Chicago: University of Chicago Press.

Tuathail, G. (1996) *Critical Geopolitics: The Politics of Writing Global Space*. London: Routledge.

Urry, J. (2003) *Global Complexity*. Cambridge: Polity.

Walker, R. B. J. (2010) *After the Globe, Before the World*. London: Routledge.

Wall, T. and Monahan, T. (2011) 'Surveillance and Violence From Afar: the Politics of Drones and Liminal Security Scapes.' *Theoretical Criminality* 15(3): 239-254.

Widder, N. (2008) *Reflections on Time and Politics*. Pennsylvania: Pennsylvania University Press.

Contributors

Shahzad Bashir is the Lysbeth Warren Anderson Professor in Islamic Studies at Stanford University and an Andrew Carnegie Fellow during the period 2015-16. (Website: http://web.stanford.edu/~sbashir/)

Kevin K. Birth is a professor of anthropology at Queens College of the City University of New York. He studies cultural concepts of time in relationship to cognition, and has conducted ethnographic research in Trinidad and on the current leap second controversy. His publications and presentations cover a wide-ranging array of topics including chronobiology and globalisation, comparative calendars, timekeeping in Roman Britain, culture and memory, cognitive neuroscience, early modern clocks, and ideas about roosters in the Middle Ages. He is the author of three books: *Any Time is Trinidad Time* (University Press of Florida), *Bacchanalian Sentiments* (Duke University Press), and most recently *Objects of Time* (Palgrave Macmillan).

Valerie Bryson is Emerita Professor of Politics at the University of Huddersfield. Her research interests focus on the overlapping areas of feminist political theory, women and politics, and the politics of time. Her most recent publication is *Feminist Political Theory*, Third edition (Palgrave Macmillan, 2016). Other books include *Gender and the Politics of Time: Feminist theory and contemporary debates* (The Policy Press, 2007), *Redefining Social Justice: New Labour rhetoric and reality,* with P. Fisher, eds. (Manchester University Press, 2011); *Sexuality, Gender and Power, Intersectional and Transnational Perspectives, with* A. Jonasdottir and K. Jones, (eds.), (Routledge, 2011).

Kathryn Marie Fisher is Assistant Professor of International Security Studies at National Defense University's College of International Security Affairs, USA. She has previously taught at Ohio University, USA. Her publications include work in *Critical Studies on Terrorism*, *Critical Perspectives on Counter-Terrorism* and the recent book *Security, Identity, and British Counterterrorism Policy.*

Robert Hassan is Associate Professor of Culture and Communication at the University of Melbourne. He writes at the intersections of politics, media and temporality. He is the author of numerous journal articles, magazine essays and book chapters. He is the author of *Age of Distraction* (2012) and *Philosophies of Media* (Routledge 2016).

Caroline Holmqvist is Research Fellow at the Swedish Institute of International Affairs, Senior Lecturer in War Studies at the Swedish Defence University and Associate Researcher at Université Libre de Bruxelles. She is co-investigator of the project 'Time and Discourses of Global Politics', funded by the Swedish Research Council, 2014-2017. She has published *Policing Wars: On Military Intervention in the Twenty-First Century* (Palgrave Macmillan, 2014), *War, Police and Assemblages of Intervention* (Eds. with Jan Bachmann and Colleen Bell) (Routledge, 2014) and *The Character of War in the 21st Century* (Eds. with Christopher Coker) (Routledge, 2010), as well as articles in *Millennium Journal of International Studies* and *Cambridge Review of International Affairs*.

Kimberly Hutchings is Professor of Politics and International Relations at Queen Mary University of London. She is the author of *Kant, Critique and Politics* (1996), *International Political Theory: re-thinking ethics in a global era* (1999), *Hegel and Feminist Philosophy* (2003), *Time and World Politics: thinking the present* (2008) and *Global Ethics: an introduction* (2010).

Tim Luecke is a postdoctoral research fellow at the Mershon Center for International Security and the managing editor for the journal *International Theory: A Journal of International Politics, Law and Philosophy*. He defended his Ph.D. dissertation on the role of political generations in international politics in 2013 at the Ohio State University. He has published on the concept of generations and its applications in the *Qualitative and Multi-Method Research* and extended the findings of his research to the field of American Politics in an article for *The Forum: A Journal in Applied Research in Contemporary Politics*. Tim is currently starting a new research project in Germany on the World War II Generation and its role in post-war Germany.

Tom Lundborg is Senior Research Fellow at the Swedish Institute of International Affairs and Senior Lecturer in Political Science at the Swedish Defence University. He is co-investigator of the project 'Time and Discourses of Global Politics', funded by the Swedish Research Council, 2014-2017. His publications include *Politics of the Event: Time, Movement, Becoming* (Routledge, 2012, 2013), as well as articles in *Security Dialogue, European Journal of International Relations, Review of International Studies, Alternatives, International Political Sociology*, and *Theory & Event*.

Tim Stevens is a Teaching Fellow in the Department of Politics and International Relations, Royal Holloway, University of London. His newest book is *Cyber Security and the Politics of Time* (Cambridge University Press, 2016). He has published articles on the politics of security in *Security Dialogue, International Political Sociology* and *Contemporary Security Issues*.

Ty Solomon is a Lecturer in International Relations in the School of Social and Political Sciences at the University of Glasgow, UK. He is the author of *The Politics of Subjectivity in American Foreign Policy Discourses* (2015, University of Michigan Press), and articles in *International Studies Quarterly, European Journal of International Relations, Review of International Studies,* and other journals.

Note on Indexing

E-IR's publications do not feature indexes due to the prohibitive costs of assembling them. If you are reading this book in paperback and want to find a particular word or phrase you can do so by downloading a free PDF version of this book from the E-IR website.

View the e-book in any standard PDF reader such as Adobe Acrobat Reader (pc) or Preview (mac) and enter your search terms in the search box. You can then navigate through the search results and find what you are looking for. In practice, this method can prove much more targeted and effective than consulting an index.

If you are using apps (or devices) such as iBooks or Kindle to read our e-books, you should also find word search functionality in those.

You can find all of our e-books at: http://www.e-ir.info/publications